Preface

A number of international airline accidents resulting in more than 1,000 fatalities have been attributed, in part, to inadequate English language proficiency or limitations in intercultural awareness of pilots and air traffic controllers (ATCs). English is an official language for pilot-controller communications, mandated by the United Nations' International Civil Aviation Organization (ICAO), and also the *de facto* universal common language for all other communications for pilots-ATCs and also for maintenance technicians, cabin crew, and ground staff. ICAO has established Language Standards for pilot-ATC communication, but implementation is the responsibility of each individual Member State's Civil Aviation Authority. Consequently, although various agencies, programs, and airline companies have produced studies, tests, and language training materials internally, many of these are proprietary, unavailable to other potential users, and limited in their scope and comprehensive understanding of linguistic and intercultural contextual factors.

There is a growing industry awareness of the need for standards and training support for other aspects of English in aviation. In 2018–2019, two international accidents occurring within six months of each other and involving the Boeing 737–800/900 MAX series of aircraft generated intense public and media interest. What is a fact and should be valuable information to the public is that most aircraft operational and safety manuals, including safety bulletins, are published in English. Globally, many pilots and maintenance workers who rely on English language documentation use English as a second or foreign language, and there are still no consistent international reading proficiency standards. In addition, initial and recurrent flight training is very often conducted in English to international technicians, pilots, or cadets, and it has not been uncommon for flight crew to bring an interpreter to ground or flight training sessions. The awareness of the possible subtle influences of language on aviation safety, as discussed in this book, can support accident investigations in general.

This book highlights the work emerging from a collaborative program between applied linguists at Georgia State University (GSU), Atlanta, GA, USA, and aviation human factors researchers and pilots and flight instructors at Embry-Riddle Aeronautical University (ERAU or heretofore, Embry-Riddle),

Daytona Beach, FL, USA. We introduce the critical role of English in aviation in a variety of contexts and the many ways it is—or should be—factored into global and national policies and practices that impact training and language assessment competencies of various stakeholders, including international pilots, US-based ATCs, flight and ground staff, and students/cadets in aeronautical institutions and flight training programs. Although we currently represent contexts and themes often prioritized by scholars in the United States, our goal is to contribute meaningfully to the global development and improvement of (1) standards of aviation English teaching, teacher training, and testing; (2) investigations of the role of language in aviation accidents; and (3) research into language as a human factor in aviation communications, customer service, and intercultural (mis)communication.

We envision the book to address directly and practically the needs of linguists and applied linguists, especially discourse analysis and assessment experts by providing insights into the political and economic contexts within which ICAO language standards operate. We offer linguistic/language-specific analyses, a range of data, and expert opinion focusing on corpus-based findings and related research, while establishing and fostering effective and efficient analysis of language needs in the aviation industry. For aviation English teachers, subject matter experts (SMEs), and materials developers, we present case studies and examples of interactional tasks, transactional exchanges, radiotelephony, and common vocabulary and phrasal patterns in aviation discourse required to communicate successfully in various roles and contexts in the aviation industry. We investigate, identify, and advocate for the level of English proficiency required for safe and efficient flight training in English and related areas not addressed by or included in ICAO Language Standards. We highlight the sharing and discussion of language and culture-related accidents and incidents for undiscovered contributory language and human factors. Sample lessons and their specific goals and objectives are also provided.

Much work remains to be done in all sectors of language in aviation, from research to developing academically well-informed aviation English teaching materials, to developing aviation operational awareness within the English language teaching and testing communities, to helping the aviation industry better understand the need for linguistic awareness. With our specialization as applied linguists and content experts in the aviation industry, we believe that we can credibly identify and advocate for "Best Practices" in aviation language/English training, teacher and trainer preparation, and language competency assessment programs. Hopefully, our work will benefit the global aviation industry, ICAO, airline companies, members of the International Civil Aviation English Association (ICAEA), and various pilot-training institutions globally, as well as, certainly and importantly, the people whom the aviation industry serves.

English in Global Aviation

ALSO AVAILABLE FROM BLOOMSBURY

Identity in Applied Linguistics Research, by Lisa McEntee-Atalianis
Second Language Acquisition in Action, by Andrea Nava and
Luciana Pedrazzini
Technolingualism, by James Pfrehm
The Bloomsbury Companion to Language Industry Studies, edited by
Erik Angelone, Maureen Ehrensberger-Dow, and Gary Massey

English in Global Aviation

Context, Research, and Pedagogy

ERIC FRIGINAL, ELIZABETH MATHEWS, AND JENNIFER ROBERTS

BLOOMSBURY ACADEMIC
LONDON • NEW YORK • OXFORD • NEW DELHI • SYDNEY

BLOOMSBURY ACADEMIC
Bloomsbury Publishing Plc
50 Bedford Square, London, WC1B 3DP, UK
1385 Broadway, New York, NY 10018, USA

BLOOMSBURY, BLOOMSBURY ACADEMIC and the Diana logo
are trademarks of Bloomsbury Publishing Plc

First published in Great Britain 2020

Copyright © Eric Friginal, Elizabeth Mathews, and Jennifer Roberts, 2020

Eric Friginal, Elizabeth Mathews, and Jennifer Roberts have asserted their right under the
Copyright, Designs and Patents Act, 1988, to be identified as Authors of this work.

For legal purposes the Acknowledgments on p. xviii constitute
an extension of this copyright page.

Cover design: Ben Anslow
Cover image: Getty Images / teekid

All rights reserved. No part of this publication may be reproduced or transmitted
in any form or by any means, electronic or mechanical, including photocopying,
recording, or any information storage or retrieval system, without prior
permission in writing from the publishers.

Bloomsbury Publishing Plc does not have any control over, or responsibility for,
any third-party websites referred to or in this book. All internet addresses given
in this book were correct at the time of going to press. The author and publisher
regret any inconvenience caused if addresses have changed or sites have ceased
to exist, but can accept no responsibility for any such changes.

A catalogue record for this book is available from the British Library.

A catalog record for this book is available from the Library of Congress.

ISBN: HB: 978-1-3500-5930-6
PB: 978-1-3500-5931-3
ePDF: 978-1-3500-5933-7
eBook: 978-1-3500-5932-0

Typeset by Integra Software Services Pvt. Ltd.
Printed and bound in Great Britain

To find out more about our authors and books visit www.bloomsbury.com
and sign up for our newsletters.

Contents

List of Illustrations vi
Preface ix
Acknowledgments xviii
List of Major Abbreviations xx

PART ONE Context 1

1 English in Global Aviation: Historical Perspectives 3

2 English Proficiency and the International Civil Aviation Organization 27

3 Language as a Human Factor in Aviation 55

PART TWO Research 79

4 English in Global Aviation: Research Perspectives 81

5 The Language of Aviation: Corpus-Based Analysis of Aviation Discourse 111

6 Pilot-Controller Communication: A Multidimensional Analysis 151

PART THREE Pedagogy 185

7 Aviation English Pedagogy: Contexts and Settings 187

8 The Development of Aviation English Programs 215

9 *Ab-Initio* Aviation English 247

Notes 273
References 275
Index 291

Illustrations

Figures

5.1 Flesch-Kincaid reading grade levels of all MRO manuals vs. descriptive and procedural texts 147

5.2 Text readability and easability scores of all MRO manuals vs. descriptive and procedural texts, all in percentages 148

6.1 Comparison of dimension scores for Dimension 1: Addressee-focused, polite, and elaborated information vs. Involved and simplified narrative 166

6.2 Comparison of dimension scores for Dimension 2: Planned, procedural talk 172

6.3 Comparison of dimension scores for Dimension 3: Managed information flow 178

7.1 Pyramid model of required infrastructure to comply with ICAO LPRs 188

8.1 Sample cloze test activity under construction 234

8.2 An example of a synchronous virtual classroom using an online platform 235

8.3 Representation of Miami International Airport (MIA) 241

Tables

2.1 Outline Process of Developing ICAO Standards 35

2.2 Summary list of ICAO Language Standards and Recommended Practices 44

3.1 Proposed Taxonomy of Communications in Aviation 73

4.1 Sample analysis of English in aviation communications from Cushing (1994) 84

4.2 Language and communication in aviation: An annotated research bibliography 95

5.1 Linguistic and contextual variables of aviation English explored using corpora 115

5.2 Sample STE-approved vocabulary and sentences 125

5.3 Top 200 most frequent words in the MRO manuals corpus 129

5.4 Top collocates of *engine* from the MRO manuals corpus 131

5.5 Top collocates of *install* from the MRO manuals corpus 132

5.6 Most frequent 4-grams of the MRO manuals corpus 135

5.7 Most frequent 5-grams and 6-grams of the MRO manuals corpus 137

5.8 Top keywords based on keyness values of the MRO manuals corpus 140

6.1 Complete list of linguistic features used in Friginal (2008, 2013a, and 2015) 157

6.2 Summary of the linguistic features of the three factors extracted from the Call Center corpus 158

6.3 Composition of corpora used in the present study 160

6.4 Sample CORPAC text with brief header information 161

7.1 ICAO language proficiency requirements holistic descriptors 195

7.2 ICAO rating scare and training/language path 207

7.3 ICAO guidelines for aviation language trainers, administrators, and materials developers 211

8.1 Concordance lines for *sir* from a corpus of aviation interactions 242

9.1 Observer Flight #1 handout from course book 271

Context, research, and pedagogy: Chapter overviews

Pilot to ATC communications are a core element of safe and effective flight. Even with current technology that alerts pilots when they are too close to one another or too close to terrain, pilots depend on ATCs to, among other things, maintain adequate separation between flights. Some of the high-profile aviation accidents heightened industry concerns as to the dangers of miscommunications due to inadequate English language proficiency, especially of international pilots. These unfortunate occurrences have resulted in major policy changes affecting the global community of travelers. For example, when the flight crew of a Kazak airline operating in Indian airspace did not appear to clearly understand the flight level instructions from the controller, they had a "level bust," aviation jargon for flying into air space that had not been assigned to them. While a level bust often results in nothing more serious than an admonition from an ATC or a written report made to a pilot's airline, in this case, the Kazak airline actually collided with a Saudi Arabian Airline Boeing 747 resulting in 349 fatalities. The very next year, India, a member of ICAO, proposed Assembly Resolution A32-16 that prompted the implementation of a range of language (English)-based initiatives globally (Mathews, 2017).

Over many decades of aviation operations, the industry has developed standardized sets of codified phraseology that has served to facilitate pilot to ATC interactions. "*Pan-pan, pan-pan*," or "*We are declaring an emergency*," are examples of standardized phraseology occurring in specific contexts in aviation. The history of the use of English as both official and *de facto lingua franca* in aviation is complex and fascinating to anyone with an interest in language, linguistics, language teaching, and second language assessment. In this book, we provide opportunities for the applied linguistics community to learn more about the many research, curricula, and teaching opportunities that must be developed and implemented in order for the aviation industry to close the many safety gaps that remain and are related to the use of English throughout the global aviation industry. Certainly, in addition, much of the content of this book is also of interest and importance and accessible to other, non-linguist professionals and stakeholders within the aviation industry and would both heighten awareness of and provide additional incentive to support many of the recommended initiatives identified herein to increase safety within the industry.

The three primary divisions of this book are **Context, Research**, and **Pedagogy** in English for aviation. Part One: Context highlights the role of English as the official language of the industry and provides an extensive review of historical perspectives. We outline the broad historical context in which ICAO language proficiency requirements exist (Chapter 2). The historical roots of English as the *lingua franca* are explained as well as the recent rapidly changing landscape of international operations in which, for example, China is hiring Korean, Brazilian, and other foreign pilots to meet their demand for pilots to support unprecedented growth in domestic and international aviation. The field of Human Factors in Aviation (Chapter 3) operates at the intersection of multiple disciplines and ICAO Document 9683, Human Factors Training Manual, identifies disciplines that comprise the field, including psychology, engineering, medicine, education, mathematics, and industrial design. In this section, we emphasize language and the field of applied linguistics as a discipline relevant to human factors in aviation, acknowledging successful intercultural communications as fundamental to aviation safety. Nearly all human factors texts and manuals identify "communication" as a critical element of safe operations, citing breakdowns in both first language and second language interactions as contributing factors in numerous accidents and incidents. Sexton and Helmreich (2000) make the case that problem-solving communications are the verbal embodiment of threat and error management for interlocutors in the cockpit, an observation aligning well with the thesis that we discuss in Part One.

Part Two: Research begins with a review of literature on pilot-ATC communication research in the domain of aviation English. One of the earlier, influential publications, Steven Cushing's (1994) *Fatal Words: Communication Clashes and Aircraft Crashes* has paved the way for applied language-based research that highlights linguistic and transactional issues as an important contributing factor to aviation accidents or to the success or failure of communication between interlocutors. In the past twenty years, there have been an increasing number of studies that focus specifically on communication difficulties due to cultural differences. Popular books such as Malcolm Gladwell's (2008) *Outliers: The Story of Success* have included features of language and culture-specific variables that contribute to miscommunication between international pilots and ATCs based in the United States. Chapter 4 synthesizes past and present research data and discusses implications and future research models and directions. Chapters 5 and 6 present the collaborative work between GSU and Embry-Riddle, which brings together "language and content" experts to produce relevant research linking the areas of sociolinguistics, English for Specific-Occupational Purposes (ESP and EOP), and corpus linguistics in exploring the language of aviation. These chapters present linguistic analyses and related implications of situating aviation English in ESP-based paradigms, the design of successful research studies, corpus data collection, and qualitative discourse analytic approaches. Corpora from flight logs, simulations, and actual interactions between pilots and controllers are now available, although on a limited basis and only through a strict approval process as stipulated and controlled by airlines and some databases. In this book, we review recent studies and highlight our new analyses of exploratory corpora of interactions between various participants in the industry, with the objective of describing the language of aviation across contexts and groups of speakers. Results have various implications in the collection and analysis of texts, the research database of interactional tasks, customer service exchanges, and survey data of needs and skills required for successfully communicating in various spaces in the aviation industry.

And finally, addressing Pedagogy (Part Three), we focus on teaching, training, and language assessment topics and strategies in aviation English. Chapter 7 presents the landscape of aviation English pedagogy and instruction, covering the range of contexts in which aviation English is taught, including commercial airline pilots, professional ATCs, maintenance technicians, ground staff, cabin crew, and flight training. We argue for more proactive involvement and leadership of language experts in developing curricula, testing instruments, and certification programs for these language learners in the aviation industry. ICAO Language Proficiency Requirements (LPRs), as relevant for pilots and ATCs, are discussed, including their influence on current teaching and testing practices. We propose to investigate and advocate for a more thorough and compelling determination establishing English proficiency as necessary for safety in various positions within aviation, including flight training in English and other related areas not addressed by ICAO LPRs. Here, we also discuss the aviation English instructor, including his/her primary responsibilities and likely roles and responsibilities as part of a robust aviation English program. Finally, suggestions for an English language teacher unfamiliar with the aviation industry are provided. Chapter 8 suggests and recommends ideas for teachers and materials developers to effectively produce language-specific aviation English training resources. A primary goal is to help develop expert language trainers with specialization in aviation English communications in their respective instructional contexts and settings. An overview of instructional methodology recommended in aviation English is provided, including Communicative Language Teaching (CLT), Task-Based Language Teaching (TBLT), and Content-Based Language Teaching (CBLT). Sample lessons are provided that utilize corpus research and data to provide learners with authentic examples of real-world aeronautical communication. Chapter 9 describes the design of three Aviation English courses that prepare non-native English speaker (NNES) flight students for success in flight training institutions. The courses use different approaches to CBLT, integrating language skills while utilizing relevant aviation content to enable students to accomplish specific language and content goals. The importance of aviation familiarity and of English language instructors working with SMEs in the field of aviation to ensure accuracy is discussed and addressed.

Georgia State University and Embry-Riddle Aeronautical University

As noted above, this book is a product of extensive collaboration between two institutions, GSU's Department of Applied Linguistics and ESL and Embry-Riddle's Applied Aviation Sciences and College of Aeronautics. The completion of this book would not have been possible without the support of our many mentors and colleagues, especially Dr. Joan Carson (ret.) of GSU and United Airlines Captain Enrique Valdes (ret.). Dr. Carson is an applied linguist and former university administrator, with wide-ranging research and publication background and profile in English as a Second Language (ESL), culture and communications, and multilingualism. Captain Valdes has flown over 24,000 flight hours, spanning a 43-year aviation career. He is rated as a captain and has flight experience on the Boeing 727, 737, 747–400, 757, 767; the Boeing 777; and on the Cessna Citation 500 and Cessna Citation 560XL. He has extensive international flight experiences, including North and South America, Europe, and Asia. As noted in many of the chapters of this book, we also worked with co-authors who are all experts or researchers in their areas of specialization and are currently based in the United States, Brazil, and China. We also recognize the role of experts such as Dr. Dominique Estival of the MARCS Institute, Western Sydney University (Australia), and our colleagues from the International Civil Aviation English Association (ICAEA) who have supported us with a platform to present research ideas and projects and to initiate further research collaborations.

The GSU Department of Applied Linguistics and ESL is a multifaceted applied linguistics department that focuses on post-secondary/adult language learning, teaching, and use. The faculty specialize in a number of sub-disciplines, including second language (L2) acquisition, L2 writing, sociolinguistics, language assessment, corpus linguistics, educational technology, and L2 teacher education. GSU, founded in 1913, is an enterprising urban research university located in downtown Atlanta, Georgia (and other Metro Atlanta areas), with an enrollment of more than 55,000 undergraduate and graduate students and over 5,100 faculty and staff. GSU is a national leader in graduating students from widely diverse backgrounds. Over 2,000 international students from more than 176 countries, representing every world region, culture, and economic profile matriculate at the university. GSU offers 250 undergraduate and graduate degree programs in more than 100 fields of study in its eight colleges and schools. The College of Arts and Sciences' Department of Applied Linguistics and ESL consists of: (1) doctoral, master's, and undergraduate programs in applied linguistics; (2) an Intensive English Program (IEP) for pre-matriculated students; (3) credit-bearing ESL courses at the undergraduate and graduate level; and (4) a large testing program that measures the academic English proficiency of students who are non-native speakers of English. Graduate students have attended aviation English training lectures and workshops and have actively developed research projects and ESP-based lessons in collaboration with ALESL faculty. Students who have graduated from this program have found work in teaching aviation English in various locations and universities, including Embry-Riddle.

Embry-Riddle's history began in 1925 in Cincinnati, Ohio, on the twenty-second anniversary of the Wright brother's historic flight at Kitty Hawk, when barnstormer pilot John Paul Riddle and entrepreneur T. Higbee Embry formed the Embry-Riddle Company. By 1939, Riddle established the Embry-Riddle School of Aviation in Miami, Florida, and trained hundreds of American and British pilots and airplane mechanics during the Second World War. Seeking more space for a residential campus, the entire school was moved to Daytona Beach, Florida, in 1965, becoming an accredited University in 1970. Since then, Embry-Riddle's Worldwide Campus opened, initially as a way to provide education at military bases, and in 1978 the university's second residential campus began in Prescott, Arizona, with only 268 students. All three campuses have grown significantly, with the Worldwide Campus now enrolling over 23,000 students at more than 135 locations, including campuses in Brazil, Germany, and Singapore, representing Central and South America, Europe, and Asia, respectively. The two residential campuses in Daytona Beach and Prescott currently have a combined enrollment of nearly 10,000 students. Embry-Riddle offers more than 100 degrees in areas, including aviation; engineering; security, intelligence, and safety; applied science; and space. The university is renowned for its flight programs, with over 90 instructional aircraft and 130 flight simulators. Students come from all over the United States and, increasingly, from all over the world to receive top-quality flight training from Embry-Riddle; graduating students achieve a 96 percent job placement. Together, students at the three campuses come from over 140 countries to study, with international students representing 10.8 percent of the undergraduate enrollment and a striking 47.6 percent of the graduate enrollment.

Research at Embry-Riddle has grown in recent years, leading to the building of a new Research Park covering 90 acres in Daytona Beach, Florida. This multidisciplinary research facility provides local students and faculty many opportunities for "hands-on" work finding solutions to real-world problems in the aviation and aerospace industry. Embry-Riddle is also poised to become a leader in the industry for virtual reality innovation and development. One example is found in Worldwide's state-of-the-art Virtual Hub, which allows students to move through crash wreckage in a simulated environment, collecting and analyzing data, emulating an aviation accident investigator. In addition, students can build unmanned aerial vehicles (UAVs) in the Aerial

Robotics Virtual Lab, testing flight capabilities and analyzing results. Research from the residential campuses highlights advancements in other exciting areas, such as safety management systems, climate change, the NetGen Air Transportation System, and aircraft design.

With respect to aviation language, research and development in this area has escalated rapidly over the past five years. With a focus on language as a human factor in aviation safety, faculty and students have been investigating aircraft accidents to determine where and how language or culture may have played a causative or contributory role. On the training side, Embry-Riddle has begun to enthusiastically reengage in aviation English testing and training programs. In the 1990s, before ICAO LPR standards existed, the Embry-Riddle Language Institute in Daytona Beach provided language training for pilots and ATCs using in-house, tailor-made curricula and materials. The 2000s saw a decline in aviation language training at Embry-Riddle due to the loss of faculty resources, but efforts and innovations in aviation English have reemerged as all three campuses have begun initiatives to meet the language needs of aviation students both internally and externally. The Language Institutes at the Daytona Beach and Prescott campuses now both offer aviation English courses for students in the Intensive English Programs (IEPs) who intend to pursue degrees in aviation. In fact, at the Daytona Beach campus, students can take the Aviation Topics course which teaches academic skills and prepares them for the rigors of the general university experience within a learning context which these students care most about: Aviation. This course is relevant not only for future pilots or ATCs, but also for aerospace engineers, maintenance technicians, or even aviation business majors.

On the other hand, courses specifically for pilots and ATCs have also been designed, offered primarily through the Worldwide campus. In response to the growing need to train non-native English-speaking pilots using English as the language of instruction, courses designed specifically to build the language skills needed for flight training have been developed. Embry-Riddle-Worldwide will offer aviation English courses in an online environment, through both asynchronous and synchronous modalities. By taking advantage of these training options in the classroom and online, students will be able to develop their aviation English skills before beginning their flight training and, thereby, reduce the likelihood of their potentially encountering difficulty in their future studies due to inadequate language proficiency. Furthermore, Embry-Riddle provides teacher training for aviation English instructors in a variety of contexts. The university is intending to meet the infrastructure needs in the aviation language field, including the English language training needed for industry personnel, the quality assessments needed to evaluate proficiency, and the language trainers and evaluators needed to implement this training and testing.

As a leader in the global aerospace and aviation industry, Embry-Riddle is well-positioned to collaborate with colleagues all over the world to provide solutions to aviation language issues in an effort to keep our skies safe.

Eric Friginal
Elizabeth Mathews
Jennifer Roberts
Fall 2019

Acknowledgments

I met Elizabeth Mathews through Joan Carson at GSU several years ago, and it started my meaningful and very rewarding foray into studying the language of global aviation! Thanks to Elizabeth and Joan for introducing me to this dynamic area of research that I immediately claimed as directly within the domain of my specialization and expertise in the field of applied linguistics. I thank Mike Cullom greatly for his invaluable support and insights on broader professional and interpersonal communications foci, including his reviews of several drafts, guiding the completion of this book to directly address practical applications for a wider audience. As always, thanks to my mentors Douglas Biber, Randi Reppen, and Mary McGroarty of Northern Arizona University (NAU) for their constant support and encouragement and for providing me with the tools and skills to thoughtfully and systematically analyze professional intercultural discourse. Much appreciation to the faculty, staff, and students of the Department of Applied Linguistics and ESL at GSU and to my master's and doctoral students who collaborated with me on this project, Justin Taylor, Genggeng Zhang, Tamanna Mostafa, and Rachelle Udell (and also Xiaoning "Gill" Guo from the Civil Aviation Flight University of China, and Ciboney Fowler of Embry-Riddle).

-Eric Friginal, Georgia State University

With deep thanks and appreciation to my mentors and good friends, Joan Carson and Enrique "Rick" Valdes, for hundreds of hours of guidance and support over many years. In memory of my very dear friend, Reverend Elizabeth Graves, I am forever grateful for the blessing of her friendship in my life. Thanks to my daughter, Sophie, and nephew, Charlie, for so much joy and love. Most of all, with endless gratitude to my beloved husband, John R. Watret, who has supported and encouraged my journey each step of the way, making this book possible for me. You make everything better for all of us.

-Elizabeth Mathews, Embry-Riddle Aeronautical University

I would first like to thank my two professional mentors, Eric Friginal and Elizabeth Mathews, for giving me the honor of co-authoring this book with them and teaching me how to merge the fields of applied linguistics and

ACKNOWLEDGMENTS

aviation so meaningfully. I also owe my sincerest gratitude to Embry-Riddle, for trusting me to explore new opportunities, develop new courses, and build aviation English programs from the ground up in a mostly unprecedented manner. Specifically, thank you to the College of Aeronautics at Embry-Riddle-Worldwide for the constant support and providing me with the freedom, flexibility, and resources needed to pursue something new.

Thank you to my friends and my family for your support and motivating me to continue by expressing that what I do is so "cool." Particularly, thank you to my brother, Tyler Roberts, for continuing to be my best friend and the one I can always count on any time, any place. Thank you to my partner Nirmit Prabhakar for proudly standing next to me during this journey, being my forever springboard for ideas and taking a genuine interest in my work and a huge role in my happiness. Most of all, I want to thank my mother, Laura Lee Webb, who passed away a few months before being able to see the publication of this, my first book. Thank you for your unfailing, unwavering support throughout my life, putting your children before yourself and teaching us the value of hard work, perseverance, and above all, the importance of helping others.

-Jennifer Roberts, Embry-Riddle Aeronautical University

Major Abbreviations

AELS	Aviation English Language Standards
AMT	Aircraft Maintenance Technician/s
ANSP	Air Navigation Service Provider/s
ATC	Air Traffic Control/ler/s
ATIS	Automatic Terminal Information System
CAA	Civil Aviation Authority/ies
CAST	Commercial Aviation Safety Team
CBI	Content-Based Instruction
CBLT	Content-Based Language Training
CFI	Certified Flight Instructor
CL	Corpus Linguistics
CLIL	Content and Language Integrated Learning
CLT	Communicative Language Teaching
CPDLC	Controller-Pilot Data Link Communications
CPL	Commercial Pilot License
CRM	Crew Resource Management
EMI	English Medium Instruction
ERAU	Embry-Riddle Aeronautical University
ESL/EFL	English as a Second/Foreign Language
ESP/EOP/EAP	English for Specific/Occupational/Academic Purposes
FAA	Federal Aviation Administration
FMS	Flight Management System
GSU	Georgia State University
HFACS	Human Factors Analysis and Classification System
IATA	International Air Transport Association
ICAEA	International Civil Aviation English Association
ICAO	International Civil Aviation Organization
ICAO LPR	ICAO Language Proficiency Requirements
IELTS	International English Language Testing System
IEP	Intensive English Program
IFALPA	International Federation of Airline Pilots' Association
IFATCA	International Federation of Air Traffic Controllers Association
LHUFT	Language as a Human Factor in Aviation
MDA	Multidimensional Analysis

MOR	Mandatory Occurrence Report/s
MRO	Maintenance, Repair, and Overhaul
NNES/NES	Non-Native English Speaker/Native English Speaker
NTSB	National Transportation Safety Board
PRICESG	Proficiency Requirements in Common English Study Group
RTF	Radiotelephony/Radiotelephony Familiarity
SAE/STE	Simplified Aviation/Technical English
SARPS	Standards and Recommended Practices
SME	Subject Matter Expert/s
SMS	Safety Management Systems
TBLT	Task-Based Language Teaching
TEM	Threat and Error Management
TOEFL	Test of English as a Foreign Language
TESOL	Teachers of English to Speakers of Other Languages
VFR	Visual Flight Rules

PART ONE

Context

1

English in Global Aviation: Historical Perspectives

1.1. Introduction: Missed opportunities

"*Pull up, Baby.*" These are some of the final words recorded on the cockpit voice recorder on an American Airlines (AA) Boeing 757 just before it slammed into the tree line at the summit of El Deluvio Mountain near Cali, Colombia (Cockpit Voice Recorder Transcript, American Airlines Flight 965, Dec. 20, 1995) (Simmon, 1998).

On December 19, 1995, the day before his plane crashed on El Deluvio, the American pilot who, with this desperate plea, urged his jet to gain the few meters needed to clear the summit had played a game of tennis with his wife. She was a flight attendant with American Airlines. He was a captain, fifty-seven years old, and in good health. Earlier that week, they had celebrated Christmas with family in New Jersey. The official accident investigation also reported that he didn't smoke and was respected by his colleagues for his good communication skills. He had flown 13,000 hours with no accidents or incidents and had made thirteen previous trips to Cali for American Airlines.

His First Officer copilot was younger and had accumulated 5,800 flying hours. He lived in Orlando, Florida, where he and his wife home-schooled their children. The afternoon before the accident, he had played basketball with his children, and that evening, the family went to a basketball game to watch their son play. He went to bed about 11:30 p.m.

Both men commuted to Miami on December 20 for the start of the trip, arriving at the airport early for dispatch planning and pre-flight briefing. Despite their good planning, their departure to Cali was delayed for more than two hours due to late-arriving baggage and passengers from a connecting flight and an air traffic control hold because of congested airport ground traffic. The flight to Cali was three hours and twelve minutes. However, the aircraft did not arrive in Cali and neither man went home to his family. They and all but 4 of the 161 passengers and crew on AA Flight 965 lost their lives in the crash on El Deluvio.

There is almost never a single cause to an aircraft accident, and AA 965's crash in Colombia is no exception. None of the details above are incidental or irrelevant to the accident investigation; even quite personal details about the pilots' families and activities in the days and hours before the accident are important to investigators as they piece together possible latent and contributory factors that may have led to the accident. In addition to flight and flight training records, accident investigators consider what personal stress a pilot may be experiencing; whether or not he or she was encountering relationship or financial problems. What kind of temperament did the pilot have? Could pilot fatigue have been a factor? How much sleep and rest did the pilot have in the two or three days prior to the accident? What is not presented in the Accident Investigation Report in any systematic manner is information confirming the first language of the pilots or air traffic controllers (ATCs), or their degree of language proficiency in the languages used for radiotelephony communications.

According to findings in the official report of the Colombia Accident Investigation Bureau, a host of complex factors contributed to this accident. Within the set of technical, operational, and organizational factors identified, the accident investigators uncovered evidence of possible language problems in the communications between the pilots and the Cali airport ATC who was the last person in communication with the American pilots.

The December 20, 1995, accident was one of several high-profile accidents in which inadequate English language proficiency or potentially misleading English phraseology has been implicated as a possible contributory factor in the chain of events leading to the accident or incident. A close review of the circumstances of the Cali crash—the findings of the accident investigators and the causal, contributory, and latent factors they uncovered—included a particular focus on the role of language. An analysis of this particular tragedy *vis-à-vis* airline industry language proficiency requirements, including those enforced at the time of the accident as well as subsequent amendments, discloses the historical role that English-based communication and the English language in general have played within the aviation industry.

1.2. The case of American Airlines 965

Redundancy and **discipline** are intentional features of aviation safety. Engineers, mechanics, pilots, and safety experts build redundancy into the aviation system so that a failure in one part of the system is compensated for by another part of the system. Pilot training is intense, focused, and frequent. Various checklists ensure that correct procedures are followed and documented. One pilot flies the aircraft (the "pilot flying") and the other monitors the flight and manages air traffic control communications (the "pilot monitoring"). When the pilot monitoring the flight takes an action on the flight deck, he or she "calls it out" to the pilot flying, who is required to verbally acknowledge and affirm the action, so that both pilots maintain high situational awareness about the status of the aircraft at all times. Below 10,000 feet, pilots are required to maintain a "sterile cockpit"; that is, no conversation is permitted other than that directly related to the operation of the aircraft.

AA 965 departed late from Miami, and the pilots expressed some concern during the flight about possible delays. A late arrival to Cali can interfere with passengers' connecting flights. The Aeronautica Civil Report included the "flight crew's ongoing efforts to expedite their approach and landing in order to avoid potential delays" as a contributing factor to the accident. At 2136:30, the approach controller offered the pilots an opportunity to shorten their approach to Cali by flying a more direct approach to the airport, to land on runway 19 rather than runway 01 as planned in their original flight plan:

> "Sir the wind is calm. are you able to [execute the] approach [to] runway one niner?" ... The captain responded, "uh yes sir, we'll need a lower altitude right away though." The approach controller then stated, "Roger. American nine six five is cleared to VOR DME approach runway one niner. Rozo number one, arrival. Report Tulua VOR"' (AA 1965 Report, 995, p. 3).

Accepting the offer required that the pilots reconfigure the aircraft for a new flight path into the Cali airport, including accomplishing the following steps and procedures (also from the same report, p. 30):

- Locate, remove from its binder, and prominently position the chart for the approach to runway.
- Review the approach chart for relevant information such as radio frequencies, headings, altitudes, distances, and missed approach procedures.
- Select and enter data from the airplane's flight management system (FMS) computers related to the new approach.
- Compare information on the VOR DME Runway 19 approach chart with approach information displayed from FMS data.
- Verify that selected radio frequencies, airplane headings, and FMS-entered data were correct.

- Recalculate airspeeds, altitudes, configurations, and other airplane control factors for selected points on the approach.
- Hasten the descent of the airplane because of the shorter distance available to the end of new runway.
- Monitor the course and descent of the airplane, while maintaining communications with air traffic control.

The accident investigators determined that the pilots did not have adequate time to complete these tasks in order to properly and safely prepare a new approach to the airport and should have discontinued the new approach. Pilots are always "in command" of the aircraft, responsible for the safety of the flight, and have the prerogative to refuse an ATC's instruction.

The next critical error occurred when the pilots entered "R" into the flight management system (FMS), to reconfigure the flight path to Rozo Non-directional Beacon (NDB). An NDB is an electronic beacon on the ground that interacts with systems on board an aircraft to guide the aircraft along a selected flight path. However, the pilots incorrectly selected "R" identifier in the FMS, and in this case, "R" represented "Romeo" NDB. The correct FMS identifier for Rozo was "RZ"; however, "R" appeared first, and the pilot entered "R." The selection of "R" directed the aircraft to "Romeo" NDB, which was in a different direction away from the intended flight path, and the aircraft began a turn away from Cali toward the Romeo NDB location, near Bogota. It is noteworthy that Romeo was incorrectly entered without the required call-out and confirming response between the two pilots:

> One of the AA 965 pilots selected a direct course to the Romeo NDB believing that it was the Rozo NDB, and upon executing the selection in the FMS permitted a turn of the airplane towards Romeo, without having verified that it was the correct selection and without having first obtained approval of the other pilot, contrary to AA's procedures. (AA 965 Report, 1995, p. 55)

The left-hand turn off the Cali track headed the plane toward higher terrain while, at the same time, the plane was in a descent mode, and because the pilots wrongly assumed they were still on track for Cali, they continued their descent. They did not initially recognize that they had entered a wrong waypoint into their FMS, but as the aircraft began a turn to the left, the pilots began to realize that they were headed off course.

Previously, in order to expedite their descent to the Cali airport, the pilots had employed the speed brakes of the aircraft to reduce their air speed. Speed brakes are mechanisms on the surface of an aircraft's wings that can be raised or lowered to increase the drag across the wing surface and slow the plane. Moments later, when the Ground Proximity Warning System sounded, alerting the pilots to the imminent collision with terrain ("*Whoop whoop. Pull up! Pull up!*"), the pilots did not appear to recall that their speed brakes were still deployed, inhibiting the climb of the aircraft when the pilots increased power to the engines to try to lift the plane over the top of the mountain. "*Pull up Baby!*" They nearly made it. The aircraft slammed into the tree tops at an altitude of 8,900 feet on the east side of the mountain and crashed just below the summit on the West side.

1.2.1. Language as a possible contributing factor

The Accident Investigation Report of the Aeronautica Civil of the Republic of Colombia lists eighteen official "findings" of the accident investigation team. None mention language use or language proficiency. The accident investigators identified four probable causes of the accident. Again, none refer to language use or language proficiency. Four contributing factors are identified; there is no reference to anything related to language use or language proficiency of interactants. One of the twenty recommendations made by the accident investigation team does, however, urge ICAO, a specialized agency of the United Nations (UN), to take measures to ensure that pilots and ATCs strictly adhere to ICAO standards of phraseology and terminology in all radio telecommunications between pilots and controllers. (Chapter 2 explores more closely ICAO and its role in setting and promoting standards for aviation safety.)

Although not at the level of an official finding or recommendation, the AA 965 Report (1995) also notes that the Cali airspace was not provided with:

- radar coverage,
- computer software to alert aircraft deviation from a safe altitude,
- computer software to enhance the radar image of a particular flight,
- "a controller who shared a native language and culture with the flight crew." (p. 49)

Accident reports typically do not mention factors that are not relevant or potentially contributory to the accident, so it is significant that the accident investigators noted a lack of a "shared native language and culture." In order to understand how this limitation might have been a factor in the AA 965 accident, three critical linguistic contexts must be considered: (1) the nature of aeronautical radiotelephony communications, that is, the voice communications between pilots and controllers, (2) the use of phraseology and plain operational English in the unfolding pilot and controller dialogue in the moments before the accident, and (3) the language policies in place for aviation communications in the decades leading up to the AA 965 accident.

1.2.2. *The nature of aeronautical radiotelephony communications*

Aeronautical radiotelephony communications is an overarching term that refers to the communications between pilots and ATCs or between pilots communicating with pilots in other aircraft via the radiotelephone. Phraseology and plain operational language represent distinct but overlapping and related uses of English within radiotelephony communications and are equally key to maintaining a safe degree of separation between aircraft. For many decades, as phraseology was developed in use, ICAO has published *standard phraseology* for civil aviation authorities to adopt into their national regulations.

ICAO Document 9835 (2010) characterizes radiotelephony communications as "highly context-dependent," requiring a professional understanding of aviation operations, procedures, and technical themes and topics. Many of the pilot and ATC communication sequences involve routine procedures, and ATCs in particular are trained to adhere closely to standard published phraseology. Depending on the local or national regulations, in some cases, pilots receive explicit phraseology training; in many cases, pilots learn how to communicate on the radio "on the job." Radiotelephony communications require knowledge of the highly restricted sub-language that is ICAO standard phraseology as well as plain, operationally related, language proficiency.

Phraseology is simply the words and phrases codified in ICAO documents that are linked to a specific phase of flight and a particular procedure. The use of specific words and phrases for particular and mostly routine situations is prescribed in several ICAO documents: Annex 10, Volume II (Aeronautical Telecommunications); Document 4444 (Air Traffic Management), and Document 9432 (Manual of Radiotelephony).

Phraseology forms the basis of a very restricted and specialized sub-language that is described by Philips (1991) as composed of a reduced vocabulary of approximately 400 words and phrases. Utterances are brief, concise, and function words (such as articles, conjunctions, and pronouns) are deleted. Syntax is reduced and vocabulary is prescribed. Phraseology depends upon prior knowledge and shared information about the airspace and routine, expected procedures (Moder & Halleck, 2009). An example of standard phraseology in the communications near Cali is this aircraft controller instruction to AA 965:

> Roger, is cleared to Cali VOR, uh, descend and maintain one, five thousand feet. Altimeter three zero zero two. (Simmon, 1998, p. 22)

In conversational English this message would be something like the one below:

> I have received and understood your message. You are cleared to fly to the Cali VOR. You are cleared to descend your aircraft to an altitude of fifteen thousand feet. Upon reaching an altitude of fifteen thousand feet, maintain that altitude until you receive further instructions. Set your altimeters to 3002 inches of pressure.

Standard phraseology not only saves time but it also reduces the greater chance for misunderstandings resulting from variability. Much emphasis is placed on the importance of pilots and ATCs' adhering to standardized ICAO phraseology. As important as phraseology and consistent adherence to it is, there has also always been a need in radiotelephony communications for what ICAO now calls plain (operational related) language. Plain language is required not only for urgent or emergency situations, but also for even routine communications; it is used to share information, clarify, confirm, query, or negotiate an issue (Doc 9835, 2010, 3.3.17).

Although the term "phraseologies" is often used to refer to the words and phrases specified for use in specific situations, it is somewhat misleading as a countable noun. ICAO documents state clearly that the words and phrases published in ICAO documents were "not intended to be exhaustive." (ICAO Doc 4444, 2007, p. 12.2.1). When a situation arises for which there is no ICAO phraseology available, ATCs are "expected to use appropriate subsidiary phraseologies which should be as clear and concise as possible and designed to avoid possible confusion by those persons using a language other than one of their national languages" (p.12.3 prior to 2008 amendments). That is, there is no complete and definitive list of *memorizable* words and phrases that would be adequate to communicate the range of situations pilots and controllers encounter, even during routine flying. When controllers or pilots need plain language, care is to be taken so that it is as clear, concise, and accurate as language can possibly be.

In fact, pilots and controllers quite often slip between standardized routine phraseology and plain, phraseology-like language by code-switching, as in the communications between the AA 965 captain and the Cali controller after the controller's offer of a straight-in approach to Cali. In the following exchange, note the occurrence of plain language:

Approach Control: "* Sir, the wind is calm. are you able to approach runway one niner?" [* is an unintelligible word.] (Transcripts 2136:31)
AA 965: "Uh yes sir, we'll need a lower altitude right away though." (Transcript 2136:40)
"The wind is calm. Are you able to approach runway one-niner?" Plain, operational, language. (ICAO 9835; 3.3.21)

Despite ICAO efforts to standardize phraseology throughout the aviation industry, there were differences in standardized phraseology in use in any particular national airspace at the time of the AA 965 accident. Most nations adopted phrases and words that ICAO published, but there were nonetheless many differences, formal and informal, including between the US Federal Aviation Administration (FAA) phraseology and ICAO phraseology. However, a review of the Cali accident makes evident that the communication problem between the Cali controller and AA 965 pilots did not involve incorrect use of phraseology. Rather, there is, instead, evidence that a lack of ATC plain operational English proficiency inhibited communications at a critical point during the flight.

1.2.3. The use of phraseology and plain English in AA 965 accident

After the accident, the controller who was last in communication with the native English-speaking American pilots admitted to accident investigators that he had suspected from the pilot communications an anomaly in their position, that they might have been off-course. He also said that if the pilots had spoken Spanish, he would probably have sought to clarify the situation. In that case, clarification of their position would have required the use of plain language beyond the standard phraseology provided in ICAO documents. In other words, the controller was apparently proficient in English *phraseology*, but, importantly, did not have adequate plain language English proficiency to communicate his concerns to, or seek to clarify their position with, the pilots of AA 965.

Even with his limited English proficiency, the controller used the phraseology available and familiar to him to repeatedly request the aircraft heading, in an apparent attempt to alert the pilots to check or confirm their position and heading.

> The controller, who was on duty at the time of the accident, in his first interview, indicated to investigators that there were no language difficulties in the communications between himself and the accident flight crew. However, in the second interview, when asked a specific question regarding his opinion about the effects the difference in native languages between the accident flight crew and approach control may have had, he stated

that he would have asked the pilots of AA 965 more detailed questions regarding the routing and the approach if the pilots had spoken Spanish. He stated that he believed that his comprehension of the pilots' transmission was satisfactory and that the pilots also understood him. The controller said that, in a non-radar environment, it was unusual for a pilot to request to fly from his or her current position to the arrival transition. This ATC also stated that the request from AA 965 to fly direct to the Tulua VOR, when the flight was 38 miles north of Cali, made no sense to him. He said that his fluency in non-aviation English was limited and he could not ask them to elaborate on the request. Rather, he restated the clearance and requested their position relative to the Cali VOR. He believed that the flying pilot's response, that AA 965 was 37 miles from Cali, suggested that perhaps the pilot had forgotten to report passing the Tulua VOR. The controller further stated that had the pilots been Spanish-speaking, he would have told them that their request made little sense, and that it was illogical and incongruent with the realities. He said that because of limitations in his command of English he was unable to convey these thoughts to the crew. (AA 965 Report, 1995, p. 23)

By his own admission, had the controller been able to communicate to the pilots in a shared language, if either of the pilots had had Spanish language proficiency, or if the controller had had English language proficiency beyond memorized phraseology, he would have sought to clarify their position with them. Had he been able to do so, that communication might have interrupted the chain of events that led to this tragedy in which 164 passengers and crew lost their lives.

In this case, nonetheless, accident investigators did not determine that a lack of English proficiency was a cause of or a contributory factor in the accident, stating instead that the controller's use of ICAO phraseology was in compliance with ICAO Standards at that time:

> Both AA's guidance and ICAO's standards made it clear that English language ability by a controller who was not a native English speaker was limited to routine aeronautical communications. Consequently, Aeronautica Civil concludes that the Cali controller neither caused nor contributed to the cause of this accident.

1.2.4. Language policies in place at the time of AA 965 accident

Prior to 2003, an ICAO Recommended Practice in Annex 10, Volume II, Chapter 5 stated that English "shall be available" to flight crews on request:

5.1.1.1 In all situations for which standard radiotelephony phraseology is specified it shall be used.

5.2.1.1.1 Recommendation.—In General, the air ground radiotelephony communications should be conducted in the language normally used by the station on the ground.

5.2.1.2.2 Recommendation.—Pending the development and adoption of a more suitable form of speech for universal use in aeronautical radiotelephony communications, the English language should be used as such and should be available, on request from any aircraft station unable to comply with 5.2.1.1.1 at all stations on the ground serving designated airports and routes used by international air services.

Paragraph 5.1.1.1 in ICAO Annex 10, Volume II specified the use of a standard phraseology but did not specify standard ICAO phraseology. The wording of paragraph 5.1.1.1 makes clear that phraseology is not specified for all situations; by implication, there will be situations for which no standard phraseology is stipulated. In these situations, communication is necessary, and by implication plain language would be required.

What language to use for situations for which no phraseology has been specified has been variously referred to as "subsidiary phraseologies" or "plain language" in ICAO documents.

Procedures for Air Navigation Services—Air Traffic Management (DOC 4444).

Chapter 12. PHRASEOLOGIES.

Paragraph 2.1 Most phraseologies contained in Section 3 of this Part show the text of a complete message without call signs. They are not intended to be exhaustive, and when circumstances differ, pilots, ATS personnel and other ground personnel will be expected to use appropriate subsidiary phraseologies which should be as clear and concise as possible and designed to avoid possible confusion by those persons using a language other than one of their national languages.

ICAO Annex 10, paragraph 5.2.1.1.1, recommended that radiotelephony communication "should be conducted in language normally used" by the air traffic control station on the ground, and paragraph 5.2.1.2.2 recommended that English should be used when a pilot does not speak the language of the station on the ground. The first part of paragraph 5.2.1.2.2, the expectation that a "more suitable form of speech for universal use in aeronautical radiotelephony communications" would be developed, simply represents the age during which Annex 10 was introduced, in 1950, shortly after the ICAO was established, or perhaps an expectation that codifying more phraseology would be adequate to ensure safe and efficient communications.

With respect to the AA 965 accident in Colombia in 1995, the pilot spoke very little Spanish, with perhaps only the few words of greeting that were recorded in the transcript as part of his repertoire: "*Feliz navidad senorita*" as they departed the Bogota control area, and "*Buenos Señor*" in his first contact with the Cali controller. It was not easy to document this because information about any language proficiency of pilots in other languages, especially at this point, was not a systematic part of the accident investigation process. Annex 10 in ICAO documents recommended, at the time of the accident, that English be used when the pilot is "unable to comply" with ICAO Recommendation that the language of the station on the ground be used for radiotelephony communications (Annex 10; 5.2.1.1.2). The flying pilot/captain's communications with the Cali controller were in English, a combination of standard phraseology, and plain, operational language.

An ICAO Standard in Annex 1, Personnel Licensing, which was in effect at the time required that controllers should have the "ability to speak" languages "nationally designated for use in air traffic control" without "accent or impediment which would adversely affect radiotelephony communications." There was no similar language requirement specifically for pilots in ICAO documents at the time, other than the general recommendation in Annex 10 that English be used for international radiotelephony communications when needed.

1.2.5. *Investigation findings*

Language. An understanding of the scope of ICAO English language standards that were in place at the time of the accident is important because the Colombian accident investigation team was indeed able to determine that inadequate plain language proficiency had played a role in the crash, in addition to the multitude of other factors. However, a limited understanding of language use in aviation communications and an incomplete understanding of ICAO standards led the team to conclude that " … ICAO's standards made it clear that English language ability by a controller who was not a native English speaker was limited to routine aeronautical communications" (AA 965, 1995, p. 50).

Arguably, that report finding was not accurate. ICAO's Standards and Recommended Practices (SARPS), *in effect at that time,* did, in fact, acknowledge a requirement for what is now called plain language proficiency in English. Although ICAO SARPSs relating to language use in aviation communications were not as clear as linguists today would require, taken together the intent that ATCs be able to communicate in English when necessary was, at least, evident. ICAO documents stated clearly that the examples of phraseology provided for use in all situations for which they were specified, as well as the lists of ICAO phraseology, were, again, not intended to be exhaustive. Controllers were required by an ICAO Standard in Annex 1, Personnel Licensing, to speak the language(s) nationally designated for use in air traffic control. Furthermore, English was the default language for communications in cases where there was no shared language available.

Because of this misunderstanding of the subtler aspects of ICAO language requirements prior to the 2008 adoption of strengthened Language Proficiency Standards, and without a more nuanced understanding of language in aviation, the accident investigators were not able to issue a specific clarion call for a higher standard of English language proficiency for non-native English speakers.

Instead, the only recommendation related to language or language awareness in the Colombian Accident Investigation Report was that ICAO needed to urge Member States to encourage their pilots and ATCs to strictly adhere to ICAO standards phraseology and terminology in all radio telecommunications between pilots and controllers. *However, in this case, a lack of phraseology was not, in fact, the cause of the controller's inability to communicate to pilots his concern that they had flown off course.* Rather, it was the absence of plain language proficiency—together with the many other factors that always contribute ultimately to the chain of events leading to an accident.

Regardless of the shape that arguments related to the pre-2003 ICAO language standards might take, that is, whether or not ICAO SARPS required that controllers have plain English language proficiency beyond the use of ICAO phraseology, it certainly appears that the AA 965 disaster is a case in which better plain language English proficiency may have permitted a controller to interrupt the unfolding chain of events and, thereby, possibly prevent the accident. It is clear from this case that both ICAO phraseology *and* plain language proficiency are required elements for safe aviation communications, given the variation in and unpredictability of events that might present during the course of a flight.

In response to the Colombian Accident Investigation Report, the US National Transportation Safety Board (NTSB), which assisted the Colombian accident investigators in the process, issued additional recommendations:

> Therefore, the Safety Board believes that the FAA should develop, with air traffic authorities of member states of ICAO, a program to enhance controllers' fluency in common English-language phrases and interaction skills sufficient to assist pilots in obtaining situational awareness about critical features of the airspace, particularly in non-radar environments.

The intent of all of these organizations, the Colombian Accident Investigation team, ICAO, and the NTSB is quite clearly to establish language guidelines that will provide the highest level of flight safety. The missing—but essential—piece so often in aviation language policy, in training, and in accident investigation is the expertise and specialist perspectives of researchers and language teachers and trainers, especially those who are practitioners in the fields of applied linguistics, intercultural communication, and Teachers of English to Speakers of Other Languages (TESOL).

Native English speech and cultural situational awareness. The Accident Investigation Report also addressed the cultural and situational awareness of the American Airline pilots:

> The captain's communications also indicate a lack of appreciation for the differences between South American airspace and that in the United States. Terrain clearance in the United States is much more likely because of the ATC surveillance available with radar coverage over most of the airspace, the integration of computer programs with radar to alert controllers to aircraft that are descending towards terrain, and the common use of the English language. (AA 965 Report, 1995, p. 39)

As important as it is for controllers to have plain language proficiency, it is equally important for pilots operating in airspace in which English is used as a dominant, second, or foreign language, that they also bring a heightened situational awareness to the cultural and possible linguistic challenges that may be present. If the pilots flying AA 965 that night had been trained to be more situationally aware of cultural issues that can also pose a threat to accurate aviation communications, they might, for example, have been alerted to and understood the controller's unusual, repeated request for heading as a signal that they needed to increase their situational awareness and respond accordingly to ensure accurate communication. ["He said that his fluency in non-aviation English was limited and he could not ask them to elaborate on the request. *Rather, he restated the clearance and requested their position.*"].

The point here is not to point fingers or cast blame on investigators operating with the tools and information they had at their disposal at that time. And it is certainly not to cast blame on controllers or pilots who, after all, do the job for which they were trained with the language tools with which they were provided. Accident investigation is not about blame; it is about a fearless and comprehensive uncovering of any and all information that may be applied to prevent future accidents. In that regard, it is important to highlight, in retrospect, the need for accident investigators, pilots, and ATCs and those responsible for their training to have a more thorough understanding of both

ICAO language requirements, a better familiarity with Language as a Human Factor or LHUFT in aviation, and, perhaps most importantly, a knowledge and understanding of the dynamics and consequences of variations of language competency among those in critical positions in the aviation industry, especially pilots and controllers.

1.3. The importance of phraseology and plain language

Encountering, managing, and supporting speakers with varied and often limited English proficiency have been somewhat accepted in the aviation industry as integral requirements for operating in an international environment. Although there are many anecdotal reports and several high-profile accidents suggesting possible, relevant communications factors, rarely have communication difficulties due to language proficiency or lack thereof been reported *systematically* as being an industry problem that might affect safety. Instead, operators and controllers have often accommodated potentially limited English speakers, with both pilots and controllers employing a range of informal accommodation strategies. ICAO professionals and many stakeholders in the aviation industry have increasingly emphasized the role of **English as the *lingua franca*** of global aviation with nuanced forms and structures unique to speakers in this domain.

Nevertheless, many international pilots and controllers have historically communicated—and continue to communicate—in English primarily or only by memorizing a set of ICAO English phraseology. Usually, in routine situations, that has been and continues to be adequate; however, clearly, at times, it is not. Normally, operations are predictable and a pilot who knows the standardized phraseology may be proficient enough to fly into, land, and depart from an airspace in which English is the common language or the *lingua franca*. Yet, memorizing phraseology has never been and is not an adequate strategy for pilot and controller communications, especially in non-routine situations such as the AA 965 accident and other similar cases.

To summarize, ICAO standards in place in the decades prior and up to 1995 were increasingly problematic, as the aviation industry grew. [See Estival, Farris, and Molesworth (2016) for a review of English use in the early development of aviation, pp. 18–20.] ICAO language standard before 2003 did not clearly specify a language proficiency level, nor was there a consistent English fluency requirement for international pilots. There was actually an ICAO standard for controllers, but during this period, it still lacked clarity or consistency across nations or communicative tasks. The requirement for plain operational language proficiency was also not clearly stated, and the standard for phraseology did not specify if it was, in fact, an "ICAO phraseology."

AA 965 was only one of several high-profile accidents that heightened global awareness of and concern about the role of language in radiotelephony communications. A 2001 review of ICAO accident database uncovered a number of accidents in which investigators found evidence of language as a possible contributing or latent factor in the chain of events that led to the accident. These include:

1976	Trident/DC-9 midair collision, Zagreb	*176 fatalities*
1977	Double B747 runway collision, Tenerife	*583 fatalities*
1990	B707 fuel exhaustion, JFK	*73 fatalities*
1995	B757 CFIT, Cali	*160 fatalities*
1996	IL76/B747 midair collision, India	*349 fatalities*
1997	MD83/Shorts 330 runway collision, Paris/CDG	*1 fatality*
2001	MD80/Citation runway collision, Milan	*122 fatalities*

Each of these accidents illustrates certain aspects of the important role that language plays in maintaining aviation safety, and highlights gaps in the aviation language policy prior to the adoption of strengthened language proficiency requirements for pilots and ATCs in 2003.

1.4. Avianca Flight 052

Avianca Flight 052 is another case study that illustrates that the communications between pilots and ATCs are as complex as are the other contributory causes of aviation accidents. As is well documented in the Avianca 052 investigation report, communication problems appear to stem not only from a failure to use prescribed phraseology appropriately, but also from the use of ambiguous plain, operational language in aviation communications. [See also Malcolm Gladwell's (2008) account of this accident in his book *Outliers*, mentioned also in Chapter 4.]

On January 25, 1990, Avianca 052 crashed into a hill in Cove Neck, New York, after making a missed approach to land at the John F. Kennedy (JFK) airport. The weather was poor, as thunderstorms in the area caused air traffic all along the East coast to back up and aircraft were put into holding patterns. ATCs were busy and under pressure to land as many flights as they safely could.

Avianca 052 was at the end of a longer-than-expected seven-hour flight from Bogota, Colombia, to the JFK airfield. They had been placed into holding patterns over Virginia and New Jersey, and a third time near JFK. After the repeated holds and delays, the aircraft was getting dangerously low on fuel. The First Officer (FO) on board Avianca 052 was the pilot responsible for communicating with air traffic control while his captain flew the aircraft. English was not the FO's first language. The captain did not appear able to understand the English communications between the FO and air traffic control. As their fuel situation worsened, the captain repeatedly instructed the FO, in Spanish, to "tell him that we are in an emergency," and then asked, "Did you tell him we don't have fuel?" Five times the captain instructed or sought to confirm that the FO had communicated their emergency fuel situation to the ATC. Although the word for "emergency" in Spanish is very similar to English—"*Emergencia*"—the FO never used the word "emergency" in his English-language communications to the ATC about their rapidly worsening fuel situation. A wind shear alert had been issued, and as the captain attempted to land, the approach was unstable, and the captain had to conduct a "missed approach" procedure and "go-around". Shortly after the missed approach, the airplane ran out of fuel and crashed into a nearby neighborhood.

In fact, in the case of Avianca 052, the NTSB determined that the primary cause of the accident was the failure of the flight crew to adequately manage their fuel situation and a failure to communicate the severity of their fuel situation to air traffic control. As always, a complex interplay of factors contributed to the chain of events that ultimately led to this crash, including poor weather, heavy air traffic in the region, backed up because of the poor weather, and longer-than-expected flight delays. The NTSB determined that the flight crew's inadequate fuel management was a causal factor. In addition, the FO failed to communicate the severity of their fuel problem to ATC:

> The airplane exhausted its fuel supply and crashed 47 minutes after the flight crew stated that there was not sufficient fuel to make it to the alternate. This occurred after the flight was vectored for an ILS approach to the destination, missed the first approach, and was unable to complete a second approach.
>
> However, at 2054:40, when AVA052 was given a 360 turn for sequencing and spacing with other arrival traffic, the flight crew should have known that they were being treated routinely and that this situation should have prompted them to question the clearance and reiterate the criticality of their fuel condition. At that time, they could have declared an emergency. (Avianca 052, 1990, p. 52)

The NTSB attributes the inadequate communication to ATC from, basically, a lack of knowledge of FAA phraseology. The FO failed to declare an emergency or to say, "Mayday! Mayday! Mayday!" and he never used the word "emergency." He did at one point request "priority," but after they had lost two of their four engines due to fuel exhaustion: "*we just ah lost two engines and we ah need priority please*." Even at this point, the FO's communication is mitigated and rendered less urgent in affect by his use of "*please*."

There were some discussions in the NTSB report as to whether or not the term "priority" was utilized by Boeing, where the crew received training, as the terminology for the communication of fuel emergencies. Nevertheless, the NTSB found that better familiarity with FAA phraseology would have helped the crew communicate the emergency more effectively. The NTSB also recommended that the FAA should ensure that phraseology is in line with ICAO phraseology, an issue that is frequently cited as a problem for international pilots flying into the United States.

1.4.1. *Captain and FO communications*

From a linguistic and intercultural communication point of view, there is more at play in the exchanges between the captain and the FO of Avianca 052 and the FO and JFK ATC than simply inadequate use of phraseology. It is evident from the Cockpit Voice Recorder (CVR) that the captain did not speak English with any degree of proficiency and relied solely on the FO interpretations of ATC communications, a factor that the NTSB also considered. It is further evident that the FO deferred to the captain, rarely volunteering detailed and, especially, urgent information.

From the transcripts of the CVR, the captain specifically and repeatedly instructed the FO, who was managing all communications with the ATC, to communicate a fuel emergency (again, in Spanish). There were three separate instructions at the timestamps of 2124:06 (six seconds past 9:24 p.m.) in the transcript, at 2125:08, and at 2125:28:

Captain to First Officer:

2124:06	"Digale que estamos en emergencia"	(Tell him that we are in an emergency)
2124:17	"Que dijo?" (What did he say?)	
2124:22	"Digale que estamos en emergencia."	
2124:26	"Ya le dijo?" (Did you already tell him?)	
2125:08	"Digale que no tenemos combustible."	(Tell him that we don't have fuel.)
2125:28	"Ya le dijiste que no tenemos combustible."	(Did you already tell him that we don't have fuel?)

The FO replied in the affirmative each time. However, the FO communications to the ATC never specifically included any degree of urgency as was being repeatedly expressed by the captain. This fact raises language use as an issue of interest to accident investigators. The FO's turns directed to the ATC appeared to minimize the actual seriousness of their fuel exhaustion, negating the clear emergency nature and intent of the captain's repeated messages. Listening to the recordings as a discourse analyst would do, one would likely conclude that the FO's tone of voice, volume, and use of direct language most probably did not indicate a definite emergency or imminent danger. In this case, pronunciation, especially in using suprasegmental features of spoken communication, could be the focus of training for pilots in similar circumstances.

The captain clearly understood the urgency of his situation, and repeatedly asked the FO to, essentially, declare an emergency and compel the ATC to give them priority to land. The captain's inability to understand, for himself, whether or not the FO had complied with his instructions certainly contributed to the captain's fatal inability to intervene and safely operate and land the aircraft.

1.4.2. FO and ATC communications

The first reported mention to the ATC of the fuel situation occurs at 2044:58 in the recordings: "I think we need priority we're passing ... " The use of "I think" in this case reduces the urgency of the FO communication requesting priority handling. The controller follows up, and asks for Avianca's alternate options. The FO replies, "It was Boston, but we can't do it now, we, we, don't, we run out of fuel now."

In this utterance, the verb tense used (*"we run* [sic] *out of fuel now,"* instead of "We have run out of fuel," or "We ran out of fuel") again failed to communicate the urgency of the situation to the ATC and to specifically request for immediate, emergency action. In addition, the hesitancy and restart (*"We ... we ... don't ... "*) complicate the syntax of the utterance on the part of the FO. With unclear tense markers, that is, an ambiguous verb tense, the controller is potentially left to which of the following Avianca intends to communicate:

: "We [will] run out of fuel, now [if we attempt our alternate.]"
: "We [are] run[ning] out of fuel now."

At 2125:08, the captain told the FO in Spanish, "Tell him that we don't have fuel." At 2125:10, the FO radioed ATC, " ... ah, we're running out of fuel" with some form of hesitancy at the start of the turn. In this instance, the use of the progressive tense indicates an action or activity that is still in progress, implying that there is still additional time. That is, the ATC could correctly infer that the crew had not run out of fuel yet. This grammatical construction "are running" may indicate, at least for fluent English speakers, that they still have fuel, and is contrary to what the captain wished to communicate. One minute later, at 2126:35, the ATC announced its intended (i.e., proposed) action, seeking confirmation from Avianca that the instructions would be acceptable to them, considering their remaining fuel supply:

> ..ah I'm gonna bring you about fifteen miles north east and then turn you back onto the approach. Is that fine with you and your fuel?

The FO replies in the affirmative, although equivocally:

> "I guess so. Thank you very much."

At 2130:32, ATC instructed Avianca to "climb and maintain three thousand." The FO replied:

> "Ah, negative sir we just running out of fuel we okay three thousand now okay."

In this utterance, the FO first rejects the instruction to climb, tries to again communicate the fuel emergency, again inadequately, and then appears to accept the instruction to climb. Linguistically, one wonders if the FO was generally switching (i.e., negative transfer) from his first language/Spanish grammar in the second language/English.

Consider: "We just running out of fuel." In Spanish, "to have just done" something can be expressed as "*Acabar de*:"

> "*Acabo de llegar en casa*." I have just arrived home.
> "*Acabamos de terminar combustible*." We have just run out of fuel.

However, in English, the word "just" has several semantic meanings (Lee, 1987). The specificatory use of "just" is used to express the completion of a very recent activity:

> "We have just had lunch."
> "We have just run out of fuel."
> "We have just lost all four engines."

The emphatic use of "just" emphasizes a point:

> "That is just what I mean."
> "That is just fine."

"That is just the right amount of fuel."

A restrictive use of "just" restricts something, perhaps time or, as in the case of Avianca, fuel remaining:

"You have just five minutes to arrive on time."
"We are just five minutes away."
"We have just five minutes of fuel remaining."

The depreciatory use of "just" minimizes something:

"I'm just kidding."
"He just has his private pilot license."
"We are just running low on fuel."

In this light, the turn by the FO to the ATC, "We just running out of fuel," may have confused the controller as to the actual danger, leaving him to guess rather than accurately comprehend what actual meaning was intended by the FO. Did he mean:

"We are just running out of fuel." (Depreciatory) Or,
"We have just run out of fuel." (Specificatory)

It is likely that the FO intended to communicate the Spanish construction "*Acabar de*," or "We have just run out of fuel." Unfortunately, his use of "just" coupled with the progressive tense in "running" indicated to the controller that it was a situation still unfolding, and, therefore, much less urgent than the situation actually was, a series of plain language errors that aviation specialist and linguist Jeremy Mell (2018, personal communication) calls a "linguistic fault line."

1.4.3. *NTSB investigation report*

The NTSB accident investigation final report for Avianca 052 focuses first on how the flight crew could have better managed communications with the ATC.

> Even on a busy frequency, the findings suggest, a subsequent inquiry of the controller, such as, "Did you receive our low fuel call to NY Center? We said that we no longer have enough fuel to make it to our alternate?" or "We are declaring an emergency," would have been more effective to

ensure immediate sequencing to the ILS approach, according to the report. (Avianca 052, 1990, p. 64)

This suggestion presupposes, however, that the flight crew, at the time, had the necessary degree of confidence in their English language proficiency, as well as a somewhat sophisticated degree of cultural metalinguistic awareness to perceive the need for clarification of the communication, a supposition apparently not borne out upon analysis of the communication between the ATC and the flight crew as well as between the captain and the FO themselves. These abilities were undoubtedly beyond the capabilities of the captain and the FO at the time, a fact that the accident investigators appear to have either not comprehended or understood or factored in to their conclusions and recommendations. In summary, Avianca 052 illustrates the complexity of aviation communications both on the flight deck and between pilots and ATCs. In this case, the FO's limited control of pragmatic and phonetic norms of English, the captain's limited use and comprehension of English, and a communicative breakdown between the FO and ATC all contributed to the crash.

1.5. Language in accident investigations: Concluding remarks

Accident investigators are committed to a thorough and unbiased review of all the evidence available. Over many decades, accident investigations from various countries, often supported by agencies based in the United States, have expanded from an examination only of mechanical and engineering failures to the inclusion of human and organizational elements as possible contributory or latent, underlying, factors that may have contributed to the chain of events leading to an accident. Investigators examine the corporate culture of the company for which the pilots work. They seek information on the pilots' health, their home life situation, including marital status and whether or not their home life is untroubled. They investigate the number of hours of rest and sleep the pilots have had in the three days prior to the accident. Predictably, they question friends and family about pilots' drinking and/or substance use habits and whether or not there were any stressors in the pilots' lives, such as possible financial problems.

What they generally do not necessarily and consistently ask or investigate and what is not documented in any systematic way in accident investigations reports are questions and conclusions about pilot and controller language use and speakers' language proficiency. There is still also a limited focus on cultural and pragmatic awareness across the board. Accident investigators are pilots, engineers, and other technical and safety experts. In most cases, they are not routinely and systematically trained as to the role of language utilization and competency in the aviation industry and the potential threats that language-related factors might present to flight safety. As most applied linguists readily understand, language is complex and task-specific, typically mediated by register (see Chapters 5 and 6). Analyzing how communications, language, language proficiency, and culture might contribute negatively or positively to the safe operation of an aircraft requires an understanding of language beyond what an accident investigator may possess.

In both the AA 965 and Avianca 052 incidents, language issues played a role, one which accident investigators recognized at some level, but which did not rise to the level of being identified as a causal or contributing factor in the Accident Investigation Report much less mitigated through effective intervention and training. When accident investigators miss the subtler effects of language use or language proficiency, the industry underestimates the possible impact and contributory effects of language problems in the accidents being investigated. Overall, failing to clearly and expertly identify language factors in aviation represents missed opportunities to alert the industry to the importance of language in aviation safety and the need for better aviation English teaching, learning, and testing programs.

Without question, effectively addressing the role and impact of global language in the aviation industry, its potential dynamic contributions to accidents and near-misses, and the development of strategies to avoid adverse outcomes all require global analysis and solutions. Linguists and aviation safety experts familiar with the great challenges of achieving and maintaining operationally safe air travel and the requisite English proficiency at the international level to do so would agree that no blame should be assigned to an individual controller or pilot whose organization placed him or her into a position for which adequate training had not been provided. It could likewise be argued that it is also not fair to place the responsibility for developing safe levels of controller and pilot English language proficiency solely on the shoulders of ICAO Member States who do not use English as a national language, such as Colombia or China. Rather, it must be the responsibility of English-as-a-first-language nations to collaborate to bring to the global aviation community the resources that are needed to help address issues of inadequate language proficiency in international civil aviation.

There is a saying within aviation safety communities that regulations are written in the blood of aircraft accident victims. AA 965 and Avianca 052 were, in some respects, missed opportunities for accident investigators to accurately identify and highlight the critical and inevitable need for English language proficiency. Language/English and communications professionals, arguably and fortunately, possess the tools that are essential to support the achievement of these objectives. The development of effective and consistently utilized aviation English teaching and learning materials; effective, standardized competency tests; and adequate provisions for training aviation English teachers and accident investigators globally may, together, provide the greatest promise for reducing the number of accidents due, in total or in part, to communications failures.

2

English Proficiency and the International Civil Aviation Organization

Authors' note

From 2001 to 2004, Elizabeth Mathews, one of the authors of this book, worked in the Air Navigation Bureau of ICAO (International Civil Aviation Organization) Secretariat as [applied] linguistics consultant during the development of ICAO Language Proficiency Requirements (LPRs), and then intermittently as a general consultant through 2010. The information presented in this chapter reviews and synthesizes published documents and reports of ICAO's English proficiency requirements for the industry, especially for international pilots and air traffic controllers.

This introduction to ICAO, ICAO Convention, and ICAO Standards is in no way intended to represent ICAO positions on the matters described. Interpreting and applying the Articles of ICAO Convention and ICAO Standards and Recommended Practices are the domain of legal experts. Rather, this review is intended to provide general context and background to the development and implementation of ICAO language proficiency requirements, especially for practitioners in applied linguistics and TESOL, and related English training programs worldwide.

ICAO Document 9835, *Manual on the Implementation of ICAO Language Proficiency Requirements* (2nd Ed.), provides guidance to ICAO Member States. This chapter is also intended for students of applied linguistics who wish to better appreciate the significance of the language requirements in the aviation industry for their own research and academic purposes.

2.1. Introduction

It is no exaggeration to claim that aviation is the most global of all industries. No other industry is quite as dependent upon international cooperation and intercultural communication as is the global civil aviation, from the level of both long-term strategic planning and everyday basic operational and safety needs. We take for granted that we can board a plane at an airport in one city, travel safely across an ocean, through multiple time zones and the airspace of several nations, and land at an airport in another country in a few short, relatively comfortable hours. We do so because we can trust that the companies that manufacture aircraft are held to strict standards of engineering and design. We believe, correctly, that commercial aircraft are maintained to high standards of safety. We know that pilots are well-trained and subject to more rigorous training requirements than most professions. We accept that pilots understand and receive timely and accurate weather information for the airspace through which we travel. We may not know that modern aircraft are equipped with multiple alerts to notify pilots if their aircraft flies too close to a mountain or the ground or another airplane, but when we understand that, we feel secure. Unless the pilot tells us that we are in a hold for air traffic control, we may not think much about the women and men who work in airport towers and air traffic centers throughout the world and who so effectively monitor and maintain safe separation between the aircraft in their airspace. When an accident happens—and despite the great publicity that a commercial aviation accident inevitably draws when one occurs, accidents are rare—we admire the search and rescue personnel who seek, first, to save lives, and then go to extraordinary lengths to recover the remains of those who have perished so that loved ones can feel a sense of closure. Above all, when we consider it, we may feel grateful for an independent, unbiased accident investigation body whose goal is to uncover all possible factors that contributed to the accident for the specific reason of continuing to make commercial aviation safe for us, the flying public.

The high degree of safety and efficiency in commercial aviation that we, by and large, take for granted is attributable, largely, to a very extensive set of international regulations and specifications that nearly every country has agreed, voluntarily, to follow and to incorporate into their own national regulations. These regulations and specifications are the Standards, Recommended Practices, and guidance materials that are developed, published, and maintained by ICAO.

2.2. The International Civil Aviation Organization

The International Civil Aviation Organization is one of fifteen specialized agencies of the United Nations System. Many UN programs and agencies are household names: the World Health Organization (WHO); the UN Educational, Scientific, and Cultural Organization (UNESCO); and the World Bank. Others are less well-known outside of their specific domains but are organizations that provide vital global services, including the International Telecommunication Union (ITU) or the World Intellectual Property Organization (WIPO). ICAO is in the latter category, relatively unknown outside of the aviation industry but key to safe and orderly international civil aviation.

ICAO is the UN's specialized agency that is responsible for promoting development of international civil (as opposed to military) aviation throughout the world. ICAO was established in 1944, when representatives of fifty-four nations were invited by the US government to a conference in Chicago, Illinois, for the purpose of planning air routes and services and gathering data to manage the development of international aviation (Mackenzie, 2010). ICAO Convention on International Civil Aviation, or the "Chicago Convention," was signed on December 7, 1944. Yet, the need for an international body to coordinate aviation regulations was recognized even earlier.

Although thousands of years transpired from the first dreams of flight to the historic human achievement of the powered, controlled, and sustained flight of the Wright Brothers on the sea dunes of Kitty Hawk, North Carolina, by that time, the quest for flight was, nevertheless, moving full throttle on several continents. Great contributions to human understanding of the mechanics and physics that make human flight possible were made by a number of brave, curious, and committed individuals, including the Brazilian aviator and national hero, Alberto Santos-Dumont. In 1913, just ten years after these early flights, the first regularly scheduled commercial airline flights were offered, with daily flights between St. Petersburg, Florida, and nearby Tampa. At the time, for those who could afford it, the twenty-minute airplane ride represented a remarkable new convenience, reducing the travel time between the two cities from what had been a two-hour ferry boat ride, an hours-long train ride, or a nearly twenty-hour trip by car (Sharp, 2018). The first international commercial service began in 1919 between Paris and London ("Milestones in International Civil Aviation," 2018).

Very early in the history of powered flight, the need for nation states to cooperate became apparent. As early as 1910, the still very young international aviation community understood that international cooperation with efforts on both sides of the Atlantic would be required. The first aviation conference was hosted by France in 1910 in Paris (Mackenzie, 2010), and the First Conference of Pan-American Aeronautics was held in 1916, in Santiago, Chile (Mackenzie, 2010; "Milestones in International Civil Aviation," 2018). In 1919, less than twenty years after the first flight, the League of Nations sponsored a Peace Conference in Paris at which the *Convention Relating to the Regulation of Aerial Navigation* was adopted. The Paris Convention also established the International Commission on Air Navigation (ICAN), by which international aviation was guided until the adoption of the Chicago Convention. Commercial transatlantic passenger flights operated by Lufthansa between Berlin and Brooklyn, New York, began in 1938.

2.2.1. Function of ICAO

ICAO makes the safety and efficiency of international civil aviation possible by fostering and managing the cooperation and collaboration of the 192 ICAO Member States in ways that are described in this chapter. Today 192 ICAO Members States have signed the Convention. At a conference to help re-establish peace after the First World War (Mackenzie, 2010), ICAO was founded in recognition of the fact that the development of aviation can help "create and preserve friendship and understanding among the nations and peoples of the world," and that principles and agreements that support the development of international civil aviation on the "basis of equality of opportunity" were needed ("Convention on International Civil Aviation," 1944). As a UN agency, the work of ICAO is today guided by members that have signed the Convention. Each Member State is given one vote, on equal status with every other ICAO member state at ICAO Assemblies. The very first Article of the Convention notes the sovereignty of every ICAO Member State:

Article I: The contracting States recognize that every State has complete and exclusive sovereignty over the airspace above its territory.

ICAO does not act unilaterally, but rather responds to majority input from ICAO Member States. ICAO's adherence to the principle of sovereignty, seeking cooperation and collaboration rather than coercion, has been described as both its "greatest strength" and its "greatest weakness" (Mackenzie, 2010, p. 396).

Some critics, including those who have expressed concern that ICAO LPRs have not yet been fully, evenly, or legitimately implemented throughout the global aviation industry, object that ICAO has no regulatory or enforcement authority (Harris, 2012). However, it is only necessary to consider the remarkable safety record of the global commercial aviation industry, with 36.6 million takeoffs and landings in 2017, to comprehend the success that is ICAO's to claim. The degree of cooperation and collaboration within ICAO Member States, including compliance with ICAO Standards and Recommended Practices (SARPS), is precisely what makes the incredibly complex global enterprise of air transport effective, efficient, affordable, and safe.

Along with many other important activities, which include conducting and publishing the results of national safety oversight audits, a core function of ICAO is to develop, publish, and maintain the Standards and Recommended Practices and Procedures for Air Navigation that ensure that international civil aviation can operate safely and efficiently. Within the structure and functioning of ICAO, there are a number of processes and programs that support global implementation of ICAO Standards and Recommended Practices, and by signing ICAO Convention, Member States agree to abide by Article 37 of the Convention, which specifies that:

Each contracting State undertakes to collaborate in securing the highest practicable degree of uniformity in regulations, standards, procedures, and organization in relation to aircraft, personnel, airways and auxiliary services in all matters in which such uniformity will facilitate and improve air navigation.

In addition, ICAO has evolved to provide a large degree of support to Member States, including conferences, seminars, guidance documents, informative websites, and other kinds of support programs, such as those provided by ICAO's Technical Co-operation Program. ICAO proactively fosters and manages collaboration with other international aviation bodies, including formal representation to ICAO from Eurocontrol, the International Federation of Airline Pilots' Association (IFALPA), the International Air Transport Association (IATA), and the International Federation of Air Traffic Controllers Association (IFATCA). From its headquarters in Montreal, Canada, ICAO encourages and supports regional cooperation through seven regional offices: Bangkok, Cairo, Dakar, Lima, Mexico City, Nairobi, and Paris. More recently and importantly, an ICAO Universal Safety Oversight Audit program was initiated to support global standardization with transparent and expert audits of the civil aviation systems of ICAO Member States ("ICAO Assembly Resolution A32-11," 1998). Referencing all of these measures to support the implementation of ICAO Standards—evidence, perhaps, of the most remarkable cooperation to be found in international aviation—Dr. Assad Kotaite, the longest-serving ICAO President stated that there is "no room for compromise when it [comes] to air safety" (Mackenzie, 2010, p. 397). Regarding safety matters, there is a high degree of sharing, collaboration, and support within the industry. There were fifty-four nations invited to the Chicago Convention, and by 2018, 192 nations had signed the Chicago Convention.

ICAO was established to help ensure "the safe and orderly" development of international civil aviation (Preamble to ICAO Convention). An early goal and consequent achievement was to establish and standardize a set of standards, practices, and procedures that could be adopted globally by all ICAO Member States. The standards and procedures of ICAO are published and maintained in nineteen Annexes to ICAO Convention as ICAO Standards and Recommended Practices.

ICAO **Standards** are procedures, policies, or requirements that are considered *necessary* to aviation safety (ICAO Council Working Paper 1290, in the 236th Report to the Council; October 20, 1952). They appear in the Annexes as "shall" statements. An example is an ICAO Standard requiring that:

Aeroplane, airship, helicopter and powered-lift pilots, air traffic controllers and aeronautical station operators shall demonstrate the ability to speak and understand the language used for radiotelephony communications to the level specified in the language proficiency requirements in Appendix 1.

ICAO Annex 1—Personnel Licensing, Language Proficiency, paragraph 1.2.9.1

ICAO **Recommended Practices** are procedures, policies, and requirements that are "recognized as *desirable* in the interest of safety, regularity, or efficiency of international air navigation" with which Member States will "endeavor to conform" ("ICAO Council Working Paper 1290," 1952). Recommended Practices appear in the Annexes as "should" statements, preceded by the note, "Recommendation."

Recommendation.—The language proficiency of aeroplane, airship, helicopter and powered-lift pilots, flight navigators required to use the radiotelephone aboard an aircraft, air traffic controllers and aeronautical station operators who demonstrate proficiency below the Expert Level (Level 6) should be formally evaluated at intervals in accordance with an individual's demonstrated proficiency level, as follows:

a) *those demonstrating language proficiency at the Operational Level (Level 4) should be evaluated at least once every three years; and*
b) *Those demonstrating language proficiency at the Extended Level (Level 5) should be evaluated at least once every six years*

(ICAO Annex 1—Personnel Licensing, Language Proficiency, paragraph 1.2.9.6).

A Member State unable to comply with an ICAO Standard must notify ICAO, and ICAO will in turn notify other Member States of the "Difference" (Article 38). ICAO invites Member States to "extend such notification to any differences from the Recommended Practices" as well, when the "notification of such differences is important for the safety of air navigation." (Foreword to ICAO Annex 10.) Also see ICAO *Document 10055 (AN/518)—Manual on Notification and Publication of Differences*.

2.2.2. Structure of ICAO

Not only ICAO Convention but also the structure and functioning of ICAO affect how and what SARPS are proposed, adopted, and implemented. ICAO has two governing bodies, the Assembly and the Council, which consists of a technical advisory body (the Air Navigation Commission [ANC]) and a working secretariat.

The Assembly. The Assembly, the largest of the governing bodies of ICAO, with one vote per Member State, meets at least once every three years. The duties of the Assembly include reviewing and approving ICAO budget, electing representatives from the Member States who will serve on ICAO Council, guiding the strategic direction and priority projects for ICAO, overseeing all work within ICAO, and approving amendments to ICAO Convention. An ICAO Assembly is an important legislative event, and the Assembly discusses and decides crucial, sensitive, oftentimes difficult global issues. However, there is also a celebratory aspect to an ICAO Assembly. Senior representatives of the many ICAO member nations, often from within the Ministry of Transport, attend with a contingent of support staff. There are many receptions and opportunities for delegates to meet and interact with their international colleagues. Some wear traditional national or local attire, to highlight and celebrate the rich cultural custom and diversity that is also a recognized and valued aspect of ICAO. The world comes together in Montreal for those few weeks, with safety, cooperation, and collaboration as dominant themes, aspects of ICAO well worth celebrating and upon which the efficacy of the organization largely depends.

The Council. ICAO Governing Council is a permanent, year-round body. Its thirty-six representatives are elected to the Council by the Assembly and often hold the status of Ambassador to ICAO. The Assembly ensures that representation on the Council is distributed among States "of chief importance" in aviation; States that make the "largest contribution" to international civil aviation; and States appointed to ensure adequate geographical representation (ICAO Convention, Article 50).

Council members can, in a sense, be considered political appointees, with the status of Ambassador to ICAO, but they are often appointed by their home governments because of their aviation or transportation expertise. The Council reports to the Assembly and has a large scope of responsibility, including overall management of the work of ICAO, and, importantly, approving amendments to ICAO SARPS in ICAO Annexes to the Convention. The Council is led by ICAO Council President, who is elected by the Assembly every three years.

The ANC is the technical body of ICAO, which advises the Council on "all information which it considers necessary and useful for the advancement of air navigation." Appointed by the Council, ANC members are technical experts with "suitable qualifications and experience in the science and practice of aeronautics" (Article 57). The nineteen members of the ANC do not represent their own Member State, but are responsible to act as independent technical advisors.

2.2.3. *Developing ICAO Standards and Recommended Practices*

A proposal to amend an ICAO Annex can originate from within ICAO, from an ICAO Member State, or from an official ICAO international organization, such as IFALPA, IATA, or IFATCA. Regardless of the origin of a proposal, the process is iterative, involved, and transparent, allowing for review and commentary at every stage, from extensive preliminary review to discussion within the ANC to further development or revision by the Secretariat or panel or study group to which the project is assigned. If the Commission agrees to send the proposed amendments forward, they are sent for review and commentary to all ICAO Contracting States and International Organizations. Comments and assessment from Contracting States and International Organizations are gathered and analyzed by the Secretariat, and the proposals are amended accordingly, before modified proposals are brought again before the ANC for final review. The ANC can recommend they be presented to the Council, or can opt to send them back to the Secretariat for further work. If the ANC recommends they be presented to the Council, the Council members have time to review and consider the final proposals before discussion in the Council, at which point they can decide to adopt or disapprove the proposals. If the Council adopts the proposed amendments to ICAO Annexes, the proposed amendments are sent to Contracting States. Finally, ICAO Member States can either (1) approve the amendments, (2) disapprove the amendments, or (3) notify ICAO of differences, that is, disagreements or suggested revisions they wish to register. The process shown in Table 2.1 is organized to develop new or amend existing Standards, which takes approximately two years from initial proposal to adoption into ICAO Annexes.

Table 2.1 Outline process of developing ICAO Standards

Step Six	Final review by ICAO Member States
Step Five	Review, and adoption (or not), in ICAO Council
Step Four	Final review in Air Navigation Commission
Step Three	Member State review and comments gathered, analyzed
Step Two	Preliminary review by Air Navigation Commission
Step One	Development of a proposal to amend an ICAO Annex

2.3. Assembly Resolution A32-16

It was, in fact, an Assembly Resolution, A32–16, adopted by the 32nd ICAO Assembly, in 1998, that launched the program that resulted in strengthened ICAO LPRs for pilots and air traffic controllers. Since at least 1997, the Commission had been addressing concerns over the role of language in radiotelephony communications. The 1977 accident in Tenerife with the loss of 583 lives had focused international attention on the importance of accurate use of phraseology and the challenges, and even threat, that language differences can pose to safe communications ("Aircraft Accident Report," 1998). As discussed in Chapter 1, the crash of Avianca 052 in 1990, with the loss of seventy-three lives, highlighted the importance of knowing and adhering to standard phraseology, as well as the risk that unclear plain language can present to pilot—air traffic control communications. The AA 965 crash in 1995, with 157 casualties, heightened awareness of the need in aviation communications for plain operational English language proficiency (McCreary et al., 1998).

Two years previously, an Indian accident investigation team had determined that inadequate English language proficiency had been a contributing factor in a midair collision of two passenger jets over India in which 349 passengers and crew perished. In brief, the pilots of a Kazakhstan Airlines aircraft failed to follow air traffic control instructions to maintain flight level 15,000 feet and collided with a Saudi Arabian Boeing 747 jumbo jet that had been assigned an altitude of 14,000 feet. A controller had cleared the Kazakh aircraft to 15,000 feet and had alerted those pilots about a Saudi Arabian aircraft that was climbing to 14,000: "*Roger. Maintain 150. Identified traffic 12 o'clock, reciprocal Saudia Boeing 747, 14 miles. Report in sight.*" The Kazakh pilot's response to the controller's alert did not indicate comprehension of the air traffic controller message; that is, the pilot's reply, which was delayed, did not correspond to the controller's prompt. The controller again warned, "*Traffic in 13 miles, level 140.*" Without any verbal response that indicated the Kazakh pilot understood those messages, the Kazakh aircraft, which had already descended to 14,500 feet, continued to descend, colliding with the Saudi Arabian aircraft at 14,000 feet.

In response to this accident, and a growing list of aviation accidents in which accident investigators determined that inadequate English language proficiency was a contributory or latent factor, India proposed ICAO Assembly Resolution A32-16: to ensure proficiency for pilots and air traffic controllers in conducting and comprehending radiotelephony communications in English ("ICAO Assembly Resolution A32-16," 1998). Resolution A-32-16 urged "the Council to direct the Air Navigation Commission to consider this matter with a high level of priority."

A32-16: Proficiency in the English language for radiotelephony communications

Whereas recent major accident investigations have indicated lack of proficiency and comprehension of the English language by flight crews and air traffic controllers alike as a contributing factor; and

Whereas to prevent such accidents, it is essential that ICAO devise ways and means to see that all the Contracting States take steps to ensure that air traffic control personnel and flight crews involved in flight operations in airspace where the use of the English language is required, are proficient in conducting and comprehending radiotelephony communications in the English language;

The Assembly:

Urges the Council to direct the Air Navigation Commission to consider this matter with a high level of priority, and complete the task of strengthening the relevant provisions of Annex 1 and Annex 10 with a view to obligating Contracting States to take steps to ensure that air traffic control personnel and flight crews involved in flight operations in airspace where the use of the English language is required, are proficient in conducting and comprehending radiotelephony communications in the English language.

The proposed resolution was passed and assigned a high degree of priority by the Assembly, heightening the responsibility for ICAO Council, the ANC, and the Secretariat to respond and allocate resources and attention to the matter. Assembly Resolution A32-16 launched and guided, but also constrained, the work of the formal ICAO Study Group that was formed to develop recommendations to ICAO in response to the resolution. For example, A32-16 refers specifically to the language proficiency required of pilots and air traffic controllers in order to conduct and comprehend radiotelephony communications. In other words, the proposed resolution targets only the oral proficiency of pilots and air traffic controllers. The Assembly Resolution did not include reading proficiency, nor did the resolution target the related but different English as a foreign language needs of students and flight instructors engaging in initial flight training in an English-speaking context. Moreover, Assembly Resolution A32-16 did not specify the English that is required when pilots undergo jet transition training or recurrent training, nor the English reading proficiency required of maintenance personnel. Finally, A32-16 did not refer to the language proficiency that pilots who speak English as a second or foreign language need in order to communicate effectively with one another on a multilingual flight deck while operating an airplane.

The full range of operational contexts in which English is required by aviation personnel includes reading proficiency, and speaking and listening proficiency, for initial, recurrent and jet transition training. Especially when pilots encounter a difficulty or an unusual situation during flight, the need to communicate increases proportionate to the complexity and unusual circumstances of the situation, as one pilot operates the aircraft, the "pilot flying," and the other pilot, the "pilot monitoring," troubleshoots the problem. In this situation, the monitoring pilot needs to provide information to both pilots by reading aloud from on-board, and nowadays electronic, operational or emergency procedure manuals. Many, indeed most, aircraft operating manuals, on-board checklists, and Quick Reference Handbooks are in English. Reading instructions aloud is not typically a feature of ESL or EFL classrooms, nor is it a feature of most aviation English programs, but it is certainly a part of safe, on-board aviation communications. These on-board operational communications are a part of what the aviation industry refers to as crew resource management and are a key component of maintaining safe flight.

The Assembly Resolution A32-16 specifically addressed pilot and air traffic controller radiotelephony communications, likely because the language issues that had previously been identified as contributory or latent factors by accident investigators at that time focused on radiotelephony communications. Subsequent research into how reading proficiency, flight deck communications, and cross-cultural communications affect aviation operational safety and efficiency is even now relatively limited. The 2003 adoption of the strengthened ICAO LPRs helped focus on the significant challenges that many airlines and air navigation service providers have encountered in their efforts to implement ICAO requirements. Nonetheless, one positive effect is that there is now increased attention focused on all aspects of language as a factor in aviation safety. Industry awareness of language as a factor in aviation safety is growing, and as awareness increases, there are calls for more regulations to support other aspects of language use in aviation. How those issues are best addressed will be considered in the following sections.

2.4. ICAO Proficiency Requirements in Common English Study Group

In recognition of the external expertise that would be required to adequately address Task ATM-9702, the ANC approved the formation of an ICAO study group to develop and present recommendations to ICAO to improve both routine and non-routine communications, including "standardized English language testing requirements" and "minimum skill level requirements," as outlined in the Task.

Following usual ICAO procedures for the formation of Study Groups, Member States and international aviation organizations with official representation to ICAO[1] were invited to provide aviation operational and applied linguistics specialists to participate in the work of the study group, referred to as the Proficiency Requirements in Common English Study Group (PRICESG). Internationally experienced airline pilots, air traffic controllers, civil aviation administrators, and applied linguists from those organizations, as well as from a geographically diverse group of Member States, participated (Doc. 9835, paragraph 1.4.2). The work of the Study Group was directed by ICAO Technical Officers from ICAO Secretariat's Air Traffic Management Bureau and supported by a linguistic consultant contracted by ICAO.

The language specialists on the PRICESG brought a combined expertise consisting of many decades of practical experience developing, teaching, and testing English in the context of the aviation industry and academic qualifications in applied linguistics, Teaching English as a Second Language (TESL), and second language acquisition. Two PRICESG members had been instrumental in the development of the Eurocontrol aviation English test for air traffic controllers, the PELA test, a test that pre-dated the current Eurocontrol ELPAC test. Another PRICESG member brought to the project more than two decades experience administering, rating, and training other testers to administer and rate aviation-specific oral proficiency interview tests to military pilots and air traffic controllers.

2.4.1. Developing ICAO Language Proficiency Requirements

ANC Task ATM-9702 called for

> a comprehensive review of existing provisions concerning all aspects of air-ground and ground-ground voice communications in international civil aviation, aimed at the identification of deficiencies and/or shortcomings. Develop ICAO provisions, as necessary, concerning: the enhancement

of communications procedures in civil aviation, in routine or non-routine communications; standardized English language testing requirements and procedures and minimum skill level requirements in the common usage of the English language utilizing procedures and requirements already existing both within and in the world outside of civil aviation.

With the support of the PRICESG, ICAO would need to accomplish two goals. The first goal was to develop a global language policy that would apply to all Member States in equal measure, both now and into the future. At the same time, the policy would need to manage the political and technical contexts and constraints that stem from both the mission and function of ICAO, which included limited resources to apply to the project. ICAO is funded by contributions to the organization paid by Member States, with a budget that does not seem to grow consistent with the growing and ever-increasing technical complexity of the aviation industry. The second goal was to develop an ICAO language policy to comply with Article 44 of ICAO Convention to "Insure that the rights of contracting States are fully respected and that every contracting State has a fair opportunity to operate international airlines," and to "Avoid discrimination between contracting States." The language requirements had to be achievable and globally applicable to any and every Member State. The policy would need to be one that could be understood and implemented by aviation operational and training managers. Finally, it would need to be a policy that did not privilege one or a few Member States over others. Yet, at the same time, it would need to be a language policy that fulfilled the mandate set by ICAO Assembly Resolution A32-16:

> … obligating Contracting States to take steps to ensure that air traffic control personnel and flight crews involved in flight operations in airspace where the use of the English language is required, are proficient in conducting and comprehending radiotelephony communications in the English language.

All of these constraints and stipulations existed within the context of increasing industry awareness of the significant safety threat related to communications issues.

In that context, in an environment of global political and ethical considerations, limited financial resources, organizational and academic constraints, set against the backdrop of urgent safety considerations, the PRICESG worked with ICAO personnel to develop a proposal for amendments to ICAO Annexes that would obligate "States to take steps to ensure" that pilots and air traffic controllers were proficient in conducting and comprehending radiotelephony communications in the English language.

Complying with the Assembly Resolution and with the task set forth in ANC Task ATM-9702 would require amendments to ICAO Annexes, a new set of ICAO Standards and Recommended Practices—The Policy—and an aviation-specific description of the English language proficiency that would help ensure that pilots and air traffic controllers would be able to communicate adequately to manage and maintain safe flight. Implementing the strengthened ICAO language proficiency requirements would require testing instruments, the development of aviation-specific teaching and learning materials, and the training of English-as-a-foreign-language teachers so that they could develop familiarity with the aviation technical contexts required of air traffic controllers and pilots. The latter projects lay outside the scope of usual ICAO activity but called for ICAO support.[2]

2.5. ICAO Language Proficiency Requirement

Four ICAO Annexes were amended to include the new ICAO Language Proficiency Requirements:

- Annex 1—Personnel Licensing;
- Annex 6—Operation of Aircraft;
- Annex 10—Aeronautical Telecommunications, Volume II—Communication Procedures including those with PANS Status; and
- Annex 11—Air Traffic Services.

Together, the amendments to ICAO Annexes 1, 6, 10, and 11 introduced Standards and Recommended Practices that:

- Require the availability of the English language for radiotelephony communications as a matter of an ICAO Standard rather than as a Recommended Practice (Annex 10);
- Require specifically that ICAO phraseology be used (Annex 10);
- Clarify the need for plain language (Annex 1);
- Introduce an aviation-specific ICAO language proficiency rating scale (Annex 1);
- Establish **ICAO Operational Level 4** as the minimum proficiency level for flight crew and air traffic controllers (Annex 1);
- Require recurrent testing for pilots and controllers who demonstrate language proficiency below ICAO Expert Level 6 (Annex 1);

- Recommend a recurrent test schedule for such pilots and controllers (Annex 1); and
- Provide for service provider oversight of personnel compliance (Annexes 6 and 11).

2.5.1. Annex 10 — Aeronautical Telecommunications

Annex 10 is the Annex most directly related to language use. Standards in Annex 10 establish the use of English for radiotelephony communications and require that both ICAO standardized phraseology and plain language be used in these communications.

> *5.1.1.1 ICAO standardized phraseology shall be used in all situations for which it has been specified. Only when standardized phraseology cannot serve an intended transmission, plain language shall be used.*
>
> Note: Detailed language proficiency requirements appear in the Appendix to Annex 1.

Annex 10, paragraph 5.1.1.1 sets out three important requirements. First, it states clearly that ICAO phraseology is required. The previous Standard had referred to "standard radiotelephony phraseology" without specifying that ICAO phraseology be used.[3] Second, the strengthened amendment requires ICAO standardized phraseology in "all situations for which it has been specified." Pilots and air traffic controllers are obligated to know and adhere to ICAO phraseology. Plain (operationally related) language is only to be used when phraseology does not suffice.

Third, paragraph 5.1.1.1 creates an ICAO Standard that operationally related plain language shall be used when phraseology is insufficient, for both air traffic controllers and pilots. Knowing and using phraseology is not sufficient, and the strengthened ICAO language proficiency requirements clarify that pilots and air traffic controllers must demonstrate proficiency in plain language.

"Plain language" in the context of English ICAO radiotelephony communications is specific to aviation operational domains and situations, such as (1) being able to communicate adequately about a navigation failure to air traffic controller and to understand the non-routine air traffic control instructions to guide an aircraft back to the airport when the aircraft has lost navigation aids, or (2) being able to discuss the runway and emergency services that an aircraft will need in order to land while the cockpit is full of smoke and the pilots cannot see their instruments. Plain language is also routinely used in non-routine but not urgent situations, such as negotiating a reduction in speed while flying along a busy air route. Plain language can be described as aviation operationally specific English—not phraseologies.

An important clarification of previously existing ICAO language provisions was the inclusion of requirements that not only is (1) phraseology to be used, but also that (2) plain (operational) language proficiency is required for safe and effective radiotelephony communications.

2.5.2. Not an English-only policy

Annex 10, Volume II also stipulates what languages can be used for radiotelephony communications.

> *5.2.1.2 Language to be used 5.2.1.2.1 The air-ground radiotelephony communications shall be conducted in the language normally used by the station on the ground or in the English language.*

Note that the language normally used by the station on the ground may not necessarily be the language of the State in which it is located. A common language may be agreed upon regionally as a requirement for stations on the ground in that region. The level of language proficiency required for aeronautical radiotelephony communications is specified in the Appendix to Annex 1.

> *5.2.1.2.2 The English language shall be available, on request from any aircraft station, at all stations on the ground serving designated airports and routes used by international air services.*

Even some industry professionals incorrectly assume that ICAO language policy is an English-only policy; however, it is not. ICAO Standard in paragraph 5.2.1.2.1 stipulates that radiotelephony communications shall be in the language "normally used by the station on the ground or in the English language." Pilots and pilot groups have pointed out the importance of a shared common language to maintaining situational awareness, a key factor in aviation safety. Situational awareness in aviation is "the accurate perception and understanding of all the factors and conditions within the four fundamental risk elements (pilot, aircraft, environment, and type of operation) that affect safety before, during, and after the flight" ("FAA Risk Management Handbook," 2009, pp. 5–8). Pilots maintain flight safety not only by communicating with air traffic controllers but also by monitoring the communications between controllers and other pilots in the immediate airspace environment.

Maintaining situational awareness is one compelling argument for an English-only policy for radiotelephony communications. However, there are a number of factors that make it impractical to require that *only* English be used for Radiotelephony/Radiotelephony Familiarity (RTF) communications. Article 1 of ICAO Convention, which ensures State sovereignty, may be argued as one factor, but there are other factors to consider, as well. It may be, for example, that a pilot and controller with a common first language will be less effective communicating with one another using English as a foreign language than if they use their shared first language. There are legitimate and significant economic, political, national, and organizational barriers to a global English-only policy for international aviation. It is useful to recall that ICAO does not act unilaterally, but develops procedures or specifications that are in accord with 192 ICAO Member States' approval. Such a policy would require not only acceptance by the ANC, the technical body of ICAO, but it would also require approval by the thirty-two members of ICAO Council, and ultimately, acceptance and endorsement by the 192 ICAO Members States, the vast majority of whom do not have English as a first or national language.

Having stated the case against an ICAO English-only policy, it is important to note that for the sake of situational awareness and safety, the aviation industry's largest pilot association, the International Federation of Airline Pilots (IFALPA) advocates that "English should be the only language spoken in the skies" ("Language Proficiency: A Pilot's Perspective," 2008, p. 28):

> It is IFALPA's [position] that there should be a single language used for all international commercial air transport operations. This language should be English. This would provide better global awareness. (IFALPA Position Paper 14POS10; 6 September 2013)

Some nations have implemented or have moved incrementally toward IFALPA's recommendation that only English be used in aviation by implementing national regulations that require that all radiotelephony communications be conducted in English, at least at certain international centers. The Civil Aviation Authority of China announced plans to implement an English-only policy for Chinese airspace (Dennis, 2015). Some countries have become accustomed to an all-English environment for aviation communications and the transition has been successful; in others, a move toward English-only has been met with resistance, for example, in France and Spain (Tallantyre, 2014) that would not surprise any student of language policy. It is difficult to obtain data identifying which ICAO Member States may have implemented an English-only policy for radiotelephony communications, and there is no ICAO requirement that they do so. Despite the application of ICAO LPRs to any language used for radiotelephony communications, the greatest testing and teaching burden so far relates to the use of English (Daley, 2000). The requirement, as a Standard, that English be available along international air routes is embodied in SARPS in Annex 10, but the level of language proficiency specified for pilots and air traffic controllers is found in ICAO Annex 1—Personnel Licensing. Table 2.2 provides a summary of ICAO Language Standards and Recommended Practices.

Table 2.2 Summary list of ICAO Language Standards and Recommended Practices

Summary of ICAO Language Standards and Recommended Practices		
Annex 10, Vol. II, Chpt. 5.2	Strengthen the requirement that the English language be available to international flights, as a Standard.	Standard
Annex 10, Vol. II, Chpt. 5.1.1.1	Clarify the requirement for the use of both plain language and ICAO phraseologies.	Standard
Annex 10, Vol. II, Chpt. 5.1.1.1	Standardize on the use of ICAO phraseologies.	Standard
Annex 1 Chpt. 1.2.9	Establish clear minimum proficiency level requirements for flight crew members and air traffic controllers.	Standard
Annex 1 Chpt. 1.2.9	Require that pilots and controllers demonstrate at least ICAO Operational Level 4.	Standard
Annex 1, Chpt 1.2.9	Require periodic re-testing for pilots and controllers who demonstrate English language proficiency below ICAO Expert Level 6.	Standard
Annex 1, Chpt 1.2.9	Provide a testing schedule to demonstrate language proficiency.	Recommended Practice
Annexes 6 and 11	Provide for service provider and operator oversight of personnel compliance.	Standard
Annex 1 Attachment A	Introduce an ICAO language proficiency rating scale applicable to native and non-native speakers: Operational Levels 4, Extended Level 5, Expert Level 6.	Standard
Annex 1 Attachment A	Pre-Elementary Level 1; Elementary Level 2; Pre-Operational Level 3 are part of ICAO Guidance material, to aid training.	

2.5.3. Annex 1—Personnel Licensing

There are three Standards and three Recommended Practices in *Annex 1— Personnel Licensing, Section 1.2.9* related to language proficiency. A Standard at paragraph 1.2.9.1 specifies that pilot and air traffic controllers must demonstrate the language used for radiotelephony communications at the proficiency level specified as ICAO Operational Level 4 in ICAO Rating Scale.[4]

> *1.2.9.1 Aeroplane, airship, helicopter and powered-lift pilots, air traffic controllers and aeronautical station operators shall demonstrate the ability to speak and understand the language used for radiotelephony communications to the level specified in the language proficiency requirements in Appendix 1.*

One Standard and two Recommended Practices specify English language proficiency requirements for other personnel, including flight engineers, balloon and glider pilots, and flight navigators. A Standard at paragraph 1.2.9.3 requires that "Flight navigators required to use the radiotelephone aboard an aircraft shall demonstrate the ability to speak and understand the language used for radiotelephony communications." However, that Standard does not specify a proficiency level. A Recommended Practice in the next paragraph, paragraph 1.2.9.4, however, states that flight navigators who are required to use the radio to communicate with air traffic control *should* demonstrate the ability to speak and understand the language used at the level specified as ICAO Operational Level 4 in ICAO Rating Scale.

It is possible, perhaps, to discern in these two similar specifications—(1) a Standard requiring language proficiency, but with no level of proficiency specified, and (2) a level of proficiency specified as a matter of a Recommended Practice— how ICAO manages the commitment to safety while at the same time maintaining an awareness of the practical challenges facing Member States. Flight navigators were instrumental on three-person flight crews for several decades but are relatively less commonly employed on modern aircraft; therefore, ICAO has a requirement that navigators demonstrate English proficiency (in the case where it is the language being used), with ICAO Operational Level 4 proficiency level recommended.

In order to ensure the maintenance of the required level of language proficiency, an ICAO Standard (paragraph 1.2.9.5) stipulates that

> *1.2.9.5 The language proficiency of aeroplane, airship, helicopter and powered-lift pilots, air traffic controllers and aeronautical station operators who demonstrate proficiency below the Expert Level (Level 6)* **shall**

be formally evaluated at intervals in accordance with an individual's demonstrated proficiency level.

A Recommended Practice at paragraph 1.2.9.6 clarifies how frequently retesting should occur: every three years or every six years.

1.2.9.6 Recommendation.— The language proficiency of aeroplane, airship, helicopter and powered-lift pilots, flight navigators required to use the radiotelephone aboard an aircraft, air traffic controllers and aeronautical station operators who demonstrate proficiency below the Expert Level (Level 6) should be formally evaluated at intervals in accordance with an individual's demonstrated proficiency level, as follows:

a) those demonstrating language proficiency at the Operational Level (Level 4) should be evaluated at least once every three years; and

b) those demonstrating language proficiency at the Extended Level (Level 5) should be evaluated at least once every six years.

A brief but important Standard in ICAO *Annex 6—Operation of Aircraft* and another in *Annex 11—Air Traffic Services* assign responsibility to airlines and air navigation service providers to ensure that their pilots and controllers comply with ICAO LPRs. Airlines and air navigation services are the "front-line" of aviation safety; ICAO Standards require that they are ultimately responsible for the adequate English language proficiency of their pilots and controllers in their employ.[5]

2.6. Evaluation of language proficiency

Annex 1 requires a specific level of language proficiency, but one of the primary limitations of ICAO language provisions prior to the 2003 adoption of strengthened ICAO language requirements was that no proficiency level was described. What did it mean to "have demonstrated a level of knowledge" in "the language or languages nationally designated for use in air traffic control and ability to speak such language or languages without accent or impediment which would adversely affect radiotelephony communications?" (ICAO Annex 1, 4.1.1.2). Without describing what constituted a minimum required level of proficiency, it was difficult to enact, train for, or enforce for such a requirement.

ICAO Rating Scale addresses this prior limitation by specifying six levels of language proficiency across six skill areas of language performance, together with a set of holistic descriptors that describe what proficient speakers can be expected to do in the context of radiotelephony communications.[6] A set of holistic descriptors establishes the kinds of communications required in the context of radiotelephony communications:

Proficient Speakers shall:
a) *Communicate effectively in voice-only (telephone/radiotelephone) and in face-to-face situations;*
b) *Communicate on common, concrete and work-related topics with accuracy and clarity;*
c) *Use appropriate communicative strategies to exchange messages and to recognize and resolve misunderstandings (e.g. to check, confirm, or clarify information) in a general or work-related context;*
d) *Handle successfully and with relative ease the linguistic challenges presented by a complication or unexpected turn of events that occurs within the context of a routine work situation or communicative task with which they are otherwise familiar; and*
e) *Use a dialect or accent which is intelligible to the aeronautical community.*

(Appendix 1 to Annex 1—Personnel Licensing)

2.6.1. Achieving compliance

It will come as no surprise to applied linguists that achieving global compliance with ICAO language proficiency requirements is an enormous, complex, multi-faceted task. There are hundreds of thousands of licensed pilots and air traffic controllers operating throughout the world, each of whom must be individually tested and confirmed as having English language speaking and listening proficiency at least to ICAO Operational Level 4. ICAO language proficiency requirements need to be rigorous enough to support successful radiotelephony communications, especially as intercultural communications in the aviation industry increase, and flexible enough to be applied throughout the entire global aviation industry, including in regions of the world with very limited resources. In this global context, ICAO Document 9835 explains that "a language proficiency rating scale may be thought of as a guide to good judgement and an important step towards harmonization of language standards to which pilots and air traffic controllers are held" (ICAO Document 9835, Chapter 4, paragraph 4.5.1). Across the aviation industry, live oral examinations are a common element of flight training; pilots are accustomed to being accessed during live, oral interactions. ICAO Document 9835 clarifies how the ICAO Rating Scale can support aviation English language testing efforts within an industry accustomed to oral exams.

The use of a rating scale requires that certain conditions be met. Firstly, a community of users must agree upon a set of criteria upon which "admission" to the community will be based. The rating scale should reflect those criteria, and the community must agree to the use of the rating scale. Secondly, a body of well-informed and experienced raters should be formed. The experience and the background of the raters must be such that they inspire trust and can gain the confidence of both the community and the candidates who wish to join the community. The raters must reflect the values of the community and understand the criteria and the context in which the criteria occur. They must also agree upon standardized procedures for the implementation of the criteria. These experienced and trained raters commit to best practice, as outlined in codes of ethics and good practice. Both the standardized procedures and compliance with codes of ethics and practice require evidence that every practical and reasonable measure has been taken to ensure test effectiveness and fairness (reliability and validity). After these conditions are met, then the rating scale is used to make informed judgements about candidates.

ICAO Document 9835: Appendix E (ed 2) and para 6.7.4 (ed 1)

ICAO Document 9835 also recommends that aviation English language assessments be evaluated by two raters: both a language specialist and an aviation operational specialist.

One common industry misunderstanding of ICAO and ICAO LPRs is that there is an "ICAO language test." In fact, ICAO neither produces nor provides any aviation English test, and providing ICAO Aviation English testing to the industry would very likely lie outside the scope of ICAO mandate. However, ICAO does provide support and guidance to Member States and to the industry in their effort to comply with ICAO Standards and Recommended Practices. Since the adoption and publication of ICAO Rating Scale in 2003, many aviation English testing programs, both commercial and non-commercial, have been developed that apply ICAO Rating Scale. Document 9835 states that aviation language testing should be a collaboration between language specialists who have acquired familiarity with aspects of aviation operational communications and aviation operational specialists (Doc. 9835, paragraph 6.3.8.3) (2004 and 2010).

It is the responsibility of the Civil Aviation Authority of ICAO Member States to identify, or develop, aviation English tests that are officially endorsed for use under their authority. Some tests are used regionally, and even have a global reach, such as Eurocontrol's English Language Proficiency for Aeronautical Communication, or ELPAC, test (www.elpac.info).

2.6.2. ICAO's support to the industry in implementing LPRs

The global aviation industry has faced significant challenges, not only in implementing aviation English language testing that is valid, reliable, and also practical, but also in providing academically well-informed aviation English language learning programs that are relevant to the target audience. In order to support the industry, ICAO published *Document 9835—Manual on the Implementation of ICAO Language Proficiency Requirements* in 2004. In the same year, ICAO held an Aviation Language Symposium, inviting Civil Aviation Authorities, airline and air navigation service provider representatives, and aviation English language teachers to ICAO headquarters in Montreal to learn about the scope and implications of the strengthened requirements. The manual and symposium were ICAO's first projects to provide direction, guidance, and support to the industry as it began to develop the infrastructure that is required to achieve full compliance with ICAO LPRs.

Since 2005, ICAO has held multiple regional workshops, seminars, and symposia in an ongoing effort to support industry implementation efforts. In 2006, ICAO published a CD-ROM of rated speech samples as an aid to testing organizations, and in collaboration with the International Civil Aviation English Association (ICAEA), ICAO published in 2010 the second Rated Speech Sample Training Aid (cfapp.icao.int). The second, augmented, edition of *Document 9835—Manual on the Implementation of ICAO Language Proficiency Requirements* includes a more detailed treatment of specific features of radiotelephony communications and more guidance, including checklists, on the implementation of ICAO LPRs. In addition, ICAO regularly sends ICAO technical officers to meetings of the ICAEA, a professional organization for aviation English teachers.

2.6.3. Subsequent ICAO assembly resolutions related to LPRs

ICAO Assembly Resolution A32-16 launched the development of strengthened ICAO LPRs. In addition to the education and awareness outreach projects embodied in the series of ICAO documents, seminars, and workshops on the LPRs, a subsequent series of ICAO Assembly Resolutions highlights the challenges entailed in complying with these requirements. The LPRs were approved by ICAO Council on March 5, 2003. Article 42 of ICAO Convention stipulates that all "standards of qualification" shall "apply within five years after the date of adoption of such standard." Since these LPRs are considered as a licensing condition for pilots and air traffic controllers, Article 42 required

Contracting States to ensure that license holders demonstrate at least ICAO Operational Level 4 English proficiency by March 5, 2008. However, the 2008 deadline proved difficult for some Member States, and subsequent Assembly Resolutions provided various measures that could be applied in situations where full compliance could not be achieved. The 36th ICAO Assembly asked Member States to extend the applicability date of the LPRs until 2011, to give States more time to develop the language teaching and testing programs that are required.

In 2007, ICAO adopted Assembly Resolution A36-11, which (1) recognized the significant challenges facing the industry to fully implement the requirements, and (2) the fact that some Member States would be unable to comply by the 2008 applicability date. A36-11 urged ICAO Council to support Member States' efforts to implement ICAO language requirements and asked States to publish their language proficiency implementation plans, most specifically "interim measures to mitigate risk, as required, for pilots, air traffic controllers and aeronautical station operators involved in international operations" (ICAO Assembly Resolution A36-11). In other words, if a State was unable to achieve full compliance with the LPRs by the required deadline, they could apply interim measures, such as ensuring that at least one flight crew member on board a flight demonstrated ICAO Operational Level 4 English proficiency.

The Assembly also urged "Contracting States to waive the permission requirement under Article 40 of the Convention" for three years after the applicability date of March 5, 2008. As previously noted in this chapter, Member States must file a written difference with ICAO if they do not incorporate an ICAO Standard into their national regulations (Article 38). So, for example, a Member State may adopt, nationally, some air traffic control phraseology that differs from ICAO published phraseology, but they must notify ICAO when they do so, and ICAO will notify other Member States.

Article 39 of ICAO Convention requires that Member States make note of the licenses of pilots or controllers who do "not satisfy in full the conditions laid down in the international standard." Article 40 requires that in such cases, permission from each individual Member States into which such pilots may operate be sought (Article 40): "No personnel having certificates or licenses so endorsed shall participate in international navigation, except with the permission of the State or States whose territory is entered." In practical terms, Articles 39 and 40 essentially disallow Member States' filing differences on licensing requirements.

In 2010, Assembly Resolution A37-10 urged Contracting States to "continue to provide ICAO with regularly updated implementation plans." The Resolution also urged Contracting States "to take a flexible approach towards States that do not yet meet the Language Proficiency Requirements, yet are making progress as evidenced in their implementation plans." Furthermore, ICAO reminded Contracting States that "[d]ecisions concerning operations should be made on a non-discriminatory basis and not be made for the purpose of gaining economic advantage." The 2013 ICAO Assembly produced Resolution A38-8, recognizing the great challenges that some Member States continue to experience and urging Member States to provide assistance to one another.

2.6.4. A way forward

As can be seen in the degree of support from ICAO and the concern repeatedly expressed in ICAO Assembly Resolutions, the implementation of ICAO language proficiency requirements has not been smooth. The difficulty of implementing and supporting a global language policy is not surprising. The inconsistent state of aviation English testing globally has been documented by language testers. A review of commercially available aviation English teaching programs illustrates that, even at this point, few have fully adopted and implemented ICAO-recommended content-based approach to aviation English language teaching that is likely to bring the best results. (See related discussions in Chapters 7, 8, and 9.) Nonetheless, the very high stakes of aviation safety make it imperative that the community of applied linguists, language testers, and aviation English teachers continue to seek greater collaboration with the aviation industry to develop more and better programs at every level and within every context of aviation English, including aviation English teacher preparation, curriculum development, and testing.

Clearly, the 2003 adoption of ICAO LPRs was a necessary and significant safety step forward for the global aviation industry. Without question, much work remains, including the development of a comprehensive language policy for the aviation industry. International aviation is growing at a staggering rate. In addition to currently licensed pilots and air traffic controllers, aircraft manufacturer Boeing estimates that 790,000 additional pilots will be needed to meet the demand in aviation in the coming two decades, double the current aviation pilot workforce ("Boeing report and technical outlook," 2018). The largest demand is in the Asia-Pacific region, which will require 260,000 new pilots to manage growth in aviation. In addition to the hundreds of thousands of additional pilots, air traffic service providers will need many thousands of additional air traffic controllers to manage the anticipated increase in flights.

It is important to note that ICAO LPRs do not address the entire range of ways in which language use affects aviation safety. The LPRs address only speaking and listening requirements for pilot and controller radiotelephony communications. In multicultural cockpits in which pilots must use English as the common language to conduct "Threat and Error Management" communications or crew resource management communications, the demands on pilot English language proficiency may be more intense than is represented by a "low" ICAO Operational Level 4 proficiency. Initial and recurrent flight training in English demands reading proficiency, as well as intensive English language communications with flight instructors. As more airplanes are being repaired, maintained, and manufactured in ESL/EFL contexts, maintenance technicians require English language reading proficiency, and an ability to communicate verbally and in writing about the repairs that are being made to the airplanes. Current ICAO language requirements do not address the language needs of maintenance technicians, who most often work with technical manuals written in English, although Boeing notes that the industry will need more than 700,000 additional maintenance workers in the next two decades. (See Chapter 7 for a more detailed discussion of this topic.)

Now that the need for aviation language policies, programs, and personnel has been more widely recognized throughout the aviation industry, a necessary strategy is for local and regional organizations and groups to collaborate, with an eye for sharing successful lessons learned with the ICAO. As aviation English programs experience success, those who are involved in these programs can share strategies via academic channels and organizations such as the ICAEA (see Chapter 4).

Article 33 of the ICAO Convention. ICAO Member States are bound by the ICAO Convention to incorporate requirements that are "*equal to or above the minimum standards*" of ICAO:

> Certificates of airworthiness and certificates of competency and licenses issued or rendered valid by the contracting State in which the aircraft is registered, shall be recognized as valid by the other contracting States, provided that the requirements under which such certificates or licenses were issued or rendered valid are equal to or above the minimum standards which may be established from time to time pursuant to this Convention.

In practice this may mean that, as the industry moves forward from the original 2003 adoption of ICAO language requirements, Member States can begin to incorporate into their aviation English testing practices, improvements based on the research that has been done since 2003. Those insights can then be incorporated into aviation language teaching programs, and shared with the industry and with the ICAO. As confidence in projects which prove successful grows, it would be much more feasible for the ICAO to incorporate such proven new strategies into ICAO guidance material than it would be to propose amendments to existing ICAO requirements. Amending ICAO Annexes is a costly, time- and resource-intensive activity. At this point in the development of the aviation English industry, a "bottom-up" approach is an appropriate and effective strategy for improving aviation English programs, including testing, curricula, and teacher training.

2.7. Concluding remarks: Global context of ICAO LPRs

The global context of ICAO LPRs and the concomitant critical safety implications make ICAO language policy especially complex, both to develop and to implement. There is no other organization except for ICAO that could foster such a global language policy for the aviation industry. Organizations such as IFALPA or IATA are responsible for other sectors of the aviation industry, in this case, for pilots and airlines respectively. IFALPA represents more than 100,000 pilots in over 100 pilot associations, and IATA represents 280 airlines in over 120 countries. IFATCA and CANSO advocate for and represent air traffic controllers and air navigation service providers. Only ICAO, a Specialized Agency of the United Nations, privileging no single nation, no particular airline, whose 194 member states are signatory to ICAO Convention and so are obligated to incorporate ICAO Standards and Recommended Practices into their own national regulations, has the reach, the audience, the moral authority, and the responsibility to enact such a global language policy. Any such policy will inevitably be bound by constraints, both from within ICAO, in terms of resources and processes, and constraints imposed upon ICAO, in terms of compliance with ICAO Convention, or recognition of, and respect for, Member State resources.

The priority at the time when the 1998 ICAO Assembly adopted Resolution A32-16 was to develop a language policy. In reviewing ICAO LPRs, consider that on a Fall day in September 1998, representatives from Member States came together to discuss how to increase aviation safety. When India proposed that ICAO develop provisions that would compel Member States to ensure that pilots and controllers demonstrate English language proficiency, Member States represented were voting on provisions that would impose upon their own nations a set of language training and language testing requirements that would require enormous investments of effort and financial resources. Even so, not only did those 194 nations, fewer of whom use English as a first or national language, vote to pass A32-16, but they also agreed to assign it a high degree of priority. This commitment to safety that ICAO initiated, fosters, and that its Member States continue to support not only highlights the remarkably success of ICAO as the key driver of safety in international civil aviation but also characterizes the aviation industry's deep commitment overall to safety, a commitment that crosses national borders, linguistic barriers, and economic self-interest.

ICAO Members States agreed to a set of regulations that they must have known would take years and a great commitment of resources to achieve. In doing so, they chose "safety" over expediency. The ICAO language proficiency requirements introduce into the aviation industry a great need for the specialized expertise of applied linguists, language testers, and TESL practitioners. In the next chapter, we will review the great safety gaps that continue to exist in the aviation industry around so many facets of language in aviation. The need for more and better aviation English teaching materials and programs, targeting a wider audience, to include maintenance technicians and ab initio flight training and air traffic control cadets does not stand to diminish but rather to increase concurrent with the tremendous growth in the industry. The industry has not yet made enough progress achieving global compliance with ICAO language proficiency requirements; the need for academic applied linguistic input is the critical next step.

3

Language as a Human Factor in Aviation

With Aline Pacheco and Angela Albritton

3.1. Introduction

As noted in the previous two chapters, ICAO LPRs and related policies address only pilot and ATC verbal communications even though communication between and among other personnel also has a role in maintaining aviation safety. Pilot and controller communication is the primary focus of public and industry attention, at least in part, because when an accident or serious incident occurs, a record of that communication is immediately available in cockpit voice recorders, ATC recorders, and transcripts, permitting linguistic analysis of communications. It is important to remember, however, that, in addition to its required utilization for international radiotelephony communications, English language speaking, listening, reading, and writing are also the *de facto* common requirement for a wide array of other aspects of aviation. Much flight training is, itself, conducted in English-speaking contexts. Many, probably most, checklists and operational or procedural manuals are published in English. Checklists are required to be read aloud during flight, during normal operations and also during an abnormal or emergency situation. Pilots and mechanics may record maintenance repairs in English. English is very often the common language on multicultural flight decks. In addition, the English language serves as the vehicle for the dissemination of much aviation safety information. Safety concepts and research findings usually appear and are published first in English. Most operational and safety awareness training, such as the evolving concepts of Crew Resource Management (CRM), Threat and Error Management (TEM), and Safety Management Systems (SMS), occurs in English.

The majority of aviation safety information originates and is delivered in English; aircraft operations manuals, safety handbooks, much CBT and web-based training, online safety videos, seminars and conference presentations, and a wide variety of safety programs and training courses are developed in English by English-speaking SMEs, targeting an English-first-language speaking audience, and, in the case of training programs, are delivered in English. While some manuals and documents are translated, much information is available in English only. Most international safety and aviation professional seminars, conferences, and meetings are conducted in English without interpretation.

At the same time as ICAO English testing requirements have been formalized in aviation, tremendous growth in emerging international aviation markets means that aviation operations and training are, and will increasingly be, occurring in multicultural and multilingual environments. In fact, the majority of cockpits can already be considered bilingual environments, as most checklists and onboard documents are in English. Even when two pilots share and speak a common language other than English for flight deck, cabin crew, passenger, and air traffic control communications, their checklists and emergency procedures manuals are, very often, in English, requiring that the pilots engage in "code-switching," that is, moving between their native, common language and English within in the same conversation and even in the middle of a sentence:

"*Tenemos que hacer el* 'checklist.'" ["We have to do the 'checklist.'"]

While ICAO language proficiency requirements are an important and necessary first step, much work remains to be completed, not only at the level of implementation of existing ICAO proficiency requirements but also in terms of policy and programming that address other aspects of language use in aviation. In order for the aviation industry to comply with ICAO language requirements and, even more importantly, to continue to develop a capacity to understand and to address language issues across the wide and growing range of aviation contexts in which language has a safety role, the industry must be able to draw on the informed support of applied linguists, language testers, and language teaching specialists. In fact, a kind of reciprocal cross-training needs to occur: aviation safety specialists need to develop a better awareness of language and language use in aviation. To do so, they must be led by applied linguists who understand the complex role of language in a complex industry. In turn, applied linguists interested in helping to improve safety in the aviation industry must acquire an understanding of aviation operational contexts and topics.

3.2. Human factors in aviation

Modern aircraft are miracles of engineering. In the nearly twelve decades since humans achieved powered, sustained, and controlled flight, improvements to air safety have been continuous and dramatic. Aeronautical engineers, mechanics, test pilots, regulators, and safety experts collaborate to ensure that aircraft are designed, built, and rigorously tested to achieve extraordinary levels of reliability and performance. Improvements in the structural, mechanical, material, and technical aspects of flight safety are the result of many inputs, including more and better machines providing more and better data about weather, navigation, and performance. Sophisticated and sensitive on-board equipment alerts pilots to threats from other aircraft, weather, or terrain. Accident investigators are able to access flight data recorders to reconstruct in detail a failed flight's last minutes to inform subsequent measures and training to avoid future accidents.

In fact, airplanes have become so structurally, aerodynamically, and mechanically safe that the number of aviation accidents that accident investigators can attribute primarily to mechanical failure has decreased significantly, and safety experts have acknowledged that most airline accidents are caused more frequently by human error than by mechanical failure (Shappell & Wiegmann, 1996). Boeing researchers report that approximately 80 percent of airplane accidents can be attributable to human error on the part of a pilot, air traffic controller, or mechanic, in contrast to the first decades of aviation when accidents caused by mechanical failure were prevalent (Rankin, 2007). As the reliability of the machines and engines improved, safety experts began to focus more closely on the human element, at first specifically how humans interact with machines, but gradually including all aspects of human performance, including, increasingly, how humans interact with other humans. ICAO quotes a former FAA Administrator to illustrate the shift in focus from solely on the aircraft and aircraft engines to include consideration of human factors: "We spent over fifty years on the hardware, which is now pretty reliable. Now it's time to work with people" ("ICAO Doc 9683," 1998, p. 1.1.5).

The field of human factors in aviation is defined by the FAA as the multidisciplinary effort to generate and compile information about human capabilities and limitations and apply that information to produce safe, comfortable, and effective human performance. Similarly, the Australian CAA states that the consideration of human factors involves "understanding humans—our behaviour and performance. Then, from an operational perspective, we apply that human factors knowledge to optimise the fit between people and the systems in which they work, to improve safety and performance" ("SMS for Aviation," 2012). A 1975 conference in Istanbul of the International Air Transport Association (IATA) is cited by Hawkins (1993) as the genesis of the formal study of human factors in aviation.

Specialists focusing on human factors in aviation use several models to describe human performance and error. One of the earliest is James Reason's (1990) "Swiss cheese" model that illustrates how errors slip through "holes" in a multi-layered series of barriers to an aviation accident. The multiple safety barriers to an accident are represented by slices of Swiss cheese; the holes in the Swiss cheese represent holes, or gaps, in procedures or in the safety system in each layer of the accident-prevention protocol. Reason, working on safety at nuclear power plants, initially identified a number of potential levels of human errors: unsafe acts, unsafe supervision, and organizational influences.

Accidents are rare, in part because the industry has developed multiple levels of safety around every flight. Aviation is a highly regulated industry. Aircraft manufacturers produce sophisticated, scrupulously engineered aircraft that are aerodynamic, strong, reliable, and efficient, if expensive. Aircraft with multiple engines can even fly and land safely with just one operational engine. There have been instances in which an aircraft lost all engines and the pilots were, nevertheless, able to glide to a safe landing, including even a Boeing 767, the well-known so-called "Gimli Glider," that ran out of fuel en route from Montreal to Ottawa, Canada ("Air Canada Boeing 767 C-GAUN Accident, Gimli, Manitoba," 1983). Current technology permits airplanes to "fly themselves," with pilots monitoring the aircraft performance. This claim is an over-simplification, to be sure, but is becoming increasingly accurate, as unmanned aerial vehicle technology continues to improve.

Yet, even with nearly perfectly functioning machines, accidents continued to occur. Increasingly, safety experts realized that human-human interactions were as important as the human-machine interface or the operating integrity of the machines alone. In the evolution of industry understanding of human factors in aviation, researchers and investigators have come to understand that the term "pilot error" is misleading; in any aviation accident or serious incident, a host of factors contribute to any negative safety outcomes, including human performance, leadership on the flight deck, organizational and bureaucratic issues, and, certainly, communications.

3.2.1. Communications in the area of human factors in aviation

Human factors specialists have consistently identified communications as an essential component of human factors research. In a report for the Swedish Civil Aviation Authority, Dahlström, Laursen, and Bergström (2008) cited the 1977 runway collision at Tenerife accident as the accident that launched "the beginning of a new era for flight safety." According to the authors, "flight safety no longer seemed to be primarily a matter of a pilot's skills in handling their planes or even of technical reliability, but that pilots' skills relating to interaction with other people were found to be at least as important."

Improving the already high levels of safety in aviation requires addressing human factors. "Communication, specifically cross-cultural communication is one human element that is receiving renewed attention" ("ICAO Doc. 9835," 2004, p. vii). A review of human factors in aviation literature reveals a limited approach to understanding language issues in aviation prior to the adoption of ICAO language proficiency requirements. Although one of the early textbooks on human factors in aviation provides a cursory reference to communications and no references at all to language or language proficiency, Wiener and Nagel (1989) acknowledged that the gap between theory and practice is wider in radio communication procedures than in any other facet of aviation. Radio communication procedures include physical or organizational procedures, such as keying the mike, knowing when to communicate what information and in what order, and avoiding "stepping" on others. Radio communication procedures rely fundamentally upon language proficiency and proficient language use.

In another early text, *Human Factors in Flight* (1993), Frank Hawkins argues that the "human factors deficiencies in verbal communication" that contributed to accidents, such as the well-known 1977 Tenerife accident, had "plagued flight operations for very many years" (p. 170). He acknowledges that "Communication is in the mainstream of Human Factors study," and that the changes to ICAO phraseology in 1984, more than a decade before the 2003 ICAO language proficiency requirements were adopted, perhaps represented a first step toward a solution. He called for specialized monitoring of human factors aspects of communication in civil aviation by those with the appropriate expertise and the authority to initiate change. Hawkins also addressed the challenges the industry faces when grappling with emerging technology, such as Controller-Pilot Data Link Communications (CPDLC). CPDLC, or "datalink," replaces pilot and air traffic controller voice communications with text messages in oceanic airspace. Hawkins notes that, in contrast to CPDLC, "Spoken language provides rapid exchange of messages and uses pronunciation and accent for clarification." While some parts of this argument are valid—spoken language does provide the opportunity for speakers to transfer information quickly—an applied linguist's perspective can assist the industry to better assess the merits and constraints of new communication technology such as CPDLS. CPDLC is, in fact, an important emerging communication technology that is rapidly being implemented in the aviation industry, and an informed discussion of the linguistic advantages and disadvantages of this new tool at the practical level and of potential, unintended consequences of greater use of datalink, and the prevention of potential adverse outcomes, merits its own specific examination and response, something beyond the scope of this current discussion. The management of language and communication in aviation issues in the Hawkins text illustrates the complexity of language and communication factors in aviation and highlights the need for the specific perspective of and collaboration with applied linguistics.

One of the most comprehensive and comprehensible representations of the importance of language to aviation communication is provided in Harry and Linda Orlady's (1999) *Human Factors in Multi-Crew Flight Operations*. In a chapter on "Basic Communication," they outline types of communication in aviation, the role of written versus voice communications, phraseology versus plain language, *inter alia*. They also acknowledge the problem of possible inadequate language proficiency in pilot-controller communications. ICAO has recognized the importance of human performance since 1986 when its Assembly adopted Resolution A26-9 on Flight Safety and Human Factors. To promote better industry understanding and awareness of human factors in aviation, ICAO published a series of digests. The first, *Human Factors Digest No. 1: Fundamental Human Factors Concepts* drew from the texts of Hawkins and of Wiener and Nagel (1989) to introduce the SHEL Model of Human Factors, a concept ICAO explains that was first introduced in 1972 by Edwards and modified by Hawkins.

ICAO SHEL model places the human at the center of the model. Four elements comprise the SHEL Model of Human Factors, with "Liveware," or the human, at the center, interacting with the other elements. "Liveware-Hardware" interaction references the ways in which the human crew interacts with the machine, the aircraft, including ergonomics, or the human-mechanics interface such as the positioning of switches, buttons, and display panels, and even the seat on which a pilot sits. "Software" in the SHEL model refers to the non-physical aspects, such as procedures, manuals and checklists, symbols, as well as computer software. Liveware-Environment interface requires consideration of the actual physical environment in which flight operations occur, including the need to protect humans from environmental elements at high altitudes. Finally, "Liveware-Liveware" interactions are concerned with, simply, human interactions.

The field of human factors in Aviation operates at the intersection of multiple disciplines (Wise & Hopkins, 2010). ICAO Document 9683, *Human Factors Training Manual* (1998) lists thirteen disciplines, ranging from psychology, engineering, medicine to education, mathematics, and industrial design (2.4.1). Document 9683 presents several sample Human Factors Checklists. In the first, a section entitled "Information Transfer Factors" lists a number of factors related to language: "language barrier; misinterpretation of oral communications, timeliness/accuracy of verbal communications." Under "Other Personnel Factors," there is a category for "Communications (phraseology, rate of speech, pronunciation, etc.)." A second checklist, based on the SHEL model, lists two communication or language-related categories: "understanding" appears under Physical factors and a section on "Oral Communication" is listed under Liveware-Liveware Interface and includes noise interference, misinterpretation, phraseology, and content and rate of speech. The third checklist designed to assist accident investigators identifying human factors, including training deficiencies, details questions regarding human performance, such as "Were any problems ever noted with the pilot's performance after he assumed the duties of this position?" None of the approximately sixty questions included questions about the pilot's first or second language proficiency or what language was being used for pilot and controller communications or for flight deck communications.

A 2002 ICAO *Human Factors Digest* (1) acknowledges that "language problems" can decrease the effectiveness of communications but does not explore the topic, except to note that "It is the task of Human Factors training to prevent communication errors. This task includes the explanation of common communication problems as well as the reinforcement of a standard of language to ensure the error-free transmission of a message and its correct interpretation" (ICAO 1.4.24). Clearly there has long been an awareness of the important role of communication and language to aviation safety. Current efforts to better understand language as a factor in aviation safety have roots in the pioneering research of linguists and cognitive scientists who began, in the 1980s, to examine communications and language in both air traffic control and flight deck communications, for example, Billings and Cheaney (1981); Goguen, Linde, and Murphy (1986); Linde (1988); Orasanu and Fischer (1992); Fischer and Orasanu (1999); and Morrow and Prinzo (1999). As early as 1986, Goguen, Linde, and Murphy were providing innovative linguistic analysis of accident transcripts to identify unsafe communication patterns in pilot English-as-a-first-language communications during emergencies.

Despite the early, promising work from linguists, cognitive scientists, and communication specialists, overall industry understanding of language and language use in aviation has not kept up with industry understanding of other aspects of human performance, nor with the spectacular growth in aviation, creating many more instances of aviation cross-cultural communications, both on the flight deck and in other operational contexts. The role of language as a factor that makes communications possible—or not—has not been given the same systematic, consistent, comprehensive review that other human elements of human performance have received. CRM is one industry application of human factors research, an approach to training and supporting flight crew in more effective decision-making through establishing and maintaining better interpersonal communications and effective leadership (Wiener, Kanki, & Helmreich, 1993).

Every aviation accident is a tragedy; a "Controlled-flight into Terrain," or CFIT, accident is particularly troubling to the industry because it represents a properly operating aircraft that crashes due to human error. The 1972 crash of an Eastern Airlines Lockheed L-1011 TriStar into the Florida Everglades near Miami illustrates the importance of crew communications, situational awareness, and use of all available resources ("NTSB IAAR 73-14," 1972). Two pilots, a flight engineer, and mechanic in the jump seat of the aircraft became focused on a malfunctioning landing gear indicator light bulb, prior to landing at Miami. Upon executing a "missed approach," they were given radar vectors from ATC to fly west, away from the airport, so that they could troubleshoot the apparent landing gear problem. They were instructed to maintain 2,000 feet. The captain instructed the First Officer to engage the autopilot. The apparent malfunctioning of the landing gear/equipment proved, instead, to be limited to a burnt-out indicator light bulb, but the three pilots and company mechanic, whose attention was focused on determining whether or not the landing gear itself was, in fact, operating properly, failed to notice that the autopilot had accidentally switched off, and the aircraft had begun to descend. By the time they noticed, it was too late to avoid impact with the ground, and the plane crashed into the Florida Everglades. The NTSB determined that the flight crew failed "to monitor the flight instruments during the final four minutes of flight, and to detect an unexpected descent soon enough to prevent impact with the ground." Furthermore, "Preoccupation with a malfunction of the nose landing gear position indicating system distracted the crew's attention from the instruments and allowed the descent to go unnoticed" (p. 32).

The accident occurred at 11:42 p.m. It was a dark and moonless night. Over the Florida Everglades, there would have been few lights to help orient the pilots visually. An air traffic controller who noticed that the aircraft had descended to 900 feet queried the pilots of Eastern Airlines 401, but he did so with what aviation-specialist linguist Charlotte Linde would describe as "mitigated speech." He asked, "Eastern, ah 401, how are things coming along out there?" The pilots responded, "Okay, we'd like to turn around and come, come back in." Just a few seconds after this communication, the pilots realized their altitude had changed. The Eastern Airlines accident is just one of several in which accident investigators determined that human factors played a contributing role ("NTSB IAAR 73-14," 1972), and it is one in which the air traffic control communications to the pilots is of linguistic interest. In part, this accident helped spur interest in better CRM.

CRM was introduced to the airline industry in the United States at a NASA conference in 1979, during a workshop on *Resource Management on the Flightdeck* (Helmreich 1994). United Airlines was the first US carrier to incorporate both classroom and "Line-oriented Flight Training" CRM into their annual and recurrent training programs (Helmreich, 1994, p. 2). The CRM training emphasizes teamwork and communications as a primary means of enhancing safety, and includes awareness and use of all available sources of information and support, including not only flight deck communications but also collaboration with air traffic control, cabin crew, and ground personnel (Wiegman & Shappel, 2001).

Another example of the importance of CRM and communications to aviation safety is the 2005 crash of a Boeing 737 operated by Helios. Shortly after Helios 522 departed from Cyprus on August 14, 2005, upon reaching 10,000 feet, the pilots were confronted with a loud aural alert. Under Standard Operating Procedures, one pilot is assigned "pilot-flying" duties, while the other will be the "pilot-monitoring" including managing ATC communications. In an abnormal situation, such as the one Helios encountered, the pilot-flying will assume ATC communications as well, so that the pilot-monitoring can troubleshoot the problem, including referencing and reading aloud from the on-board Quick Reference Handbook or other emergency procedures checklists as needed. CRM teaches pilots to collaborate to problem-solve such issues ("Accident of the aircraft 5B-DBY of Helios Airways," 2005, p. 121).

Prior to take-off, when the aircraft is on the ground, the aural alert on the Boeing 737 indicates an improper take-off configuration warning. In the air, however, after reaching 10,000 feet, the same aural alert indicates an unpressurized aircraft. The captain, most probably, spoke German as a first language; as is true with most aircraft accident investigation reports, the captain's first language is not clearly indicated in the accident investigation report, but the report does note that English was not his first language, that he was a German national, and that he "spoke with a German accent" ("Accident of the aircraft 5B-DBY of Helios Airways," 2005, p. 123). The captain radioed to the Helios ground support for assistance. According to the accident report, there were communication difficulties:

> In particular, according to some statements there were difficulties due to the fact that the Captain spoke with a German accent and could not be

understood by the British engineer. The British engineer did not confirm this, but did claim that he was also unable to understand the nature of the problem that the Captain was encountering. The Operations Dispatcher then suggested that the Cypriot First Officer be asked to talk in Greek with the Cypriot Engineer also on duty. The language difficulties prolonged resolution of the problem, while the aircraft continued to climb. Moreover, the communication difficulties could also have been compounded by the onset of the initial effects of hypoxia. (p. 123)

However, before they were able to successfully identify and resolve the problem, the pilots lost consciousness. The aircraft continued to fly on autopilot until it ran out of fuel and crashed into the Greek countryside. Accident investigators determined that, "Before hypoxia began to affect the flight crew's performance, inadequate CRM contributed to the failure to diagnose the pressurization problem." The report also determines that "language difficulties prolonged resolution of the problem, while the aircraft continued to climb" (p. 123). Although the accident investigators identified language issues in the flight, noting in particular, the difficulty that ground staff had in understanding the captain's messages about the alert, and including a few other issues related to possible language and culture issues, they did not determine that language or language proficiency was a contributory factor in the chain of events that led to the accident. The accident occurred in 2005, after ICAO language proficiency requirements were adopted by ICAO Council but before the 2008 target date when the requirements moved from having a status of Recommended Practice to Standard. In addition, by 2005, aviation English language testing had not yet become operational, or readily available.

This accident illustrates the importance of CRM, and it also indicates the importance of clear communications to effective CRM. With industry recognition and acknowledgment of the importance of aviation human factors in maintaining safety, the approach to CRM has also evolved, or expanded, into an overarching conceptual framework for identifying, planning for, and recovering from known threats to aviation safety known as Threat and Error Management or TEM (Merritt & Klinect, 2006). Encompassing CRM, a TEM approach to aviation safety helps pilots identify threats and to notice and "trap" errors (an incorrect input into the flight management system, for example). By 2006, ICAO had introduced a Standard requiring that airlines provide both initial and annual recurrent TEM training to pilots (Merritt & Klinect, 2006, p. 17).

Flin, O'Connor, and Crichton (2008) analyze communication as a non-technical skill required for safety and point out that the exchange of information is a core activity for decision-making, situation awareness, team co-ordination, and leadership. Their work does not, however, specifically address language issues as a factor. Promising new directions for better comprehension of language as a factor in aviation safety can be found in documents from ICAO's Commercial Aviation Safety Team's (CAST) "Common Taxonomy Group," the purpose of which is to "develop common taxonomies and definitions for aviation accident and incident reporting systems" ("Aviation Occurrence Categories, Definitions, and Usage Notes," 2011, p. 1). CAST publishes a Human Factors Taxonomy and categorizes human factors elements as follows: environmental, experience and knowledge, organization oversight, perceptual, physical or sensory, procedural or task performance, and psychological. "Communications" appears under the category of "procedural or task performance" factors and is defined as "Factors related to communication between crew members and other groups." The definition is supplemented by usage notes: "Communication includes but not limited to lack of communication, accuracy of communication, use of phraseology, language or accent interference, misinterpretations or misunderstandings of communication and readback issues" ("Aviation Occurrence Categories, Definitions, and Usage Notes," 2011, p. 10).

Communication factors in aviation are broad and varied. It is noteworthy and encouraging that the previous edition of the CAST Human Factors Taxonomy, published prior to the 2003 adoption of ICAO language Standards, did not include "communications" as a distinct category, but the 2011 edition not only does so but also specifically references "language." This is a positive step for advancing industry understanding of how language impacts aviation safety. ICAO CAST Taxonomy referencing language will also encourage further research. Nonetheless, while the inclusion of "communications" as a subcategory in the Cast Taxonomy is a positive step, categorizing communication factors under "Procedural or task performance" is still a too limited view of communications in aviation. Understanding full range and importance of "Communication" factors requires a separate category. In any case, in this matter, as in all issues related to language factors in aviation, input from specialists in the area of applied linguistics will support a more comprehensive and informed understanding of the importance of language in aviation.

ICAO language proficiency requirements have been introduced into the industry after a forty-year effort within the industry to promote greater awareness of human factors in aviation. Sexton and Helmreich point out, "*Problem solving communications are the verbal embodiment of threat and error management in the cockpit*" (2003). Although ICAO language proficiency requirements do not specifically apply to the English language proficiency that is required for effective pilot flight deck communications, for CRM, or for the communications required for successful TEM, they have helped focus attention specifically on the role of language across a range of aviation contexts.

3.2.2. Culture in aviation communications

Since language is heavily influenced by culture and does not exist in a vacuum, a brief introduction to how the aviation industry has approached culture in aviation will help in placing language into the context of aviation human factors. The understanding of national cultures has been referenced in numerous articles and texts in relation to a multitude of aviation topics. Helmreich (1994) provides a detailed analysis of the Avianca flight 052 accident, and Merritt's (2000) study also analyzes the interplay of national cultures in the cockpit. Strauch (2010) introduces cultural differences and socio-technical system operations, and Soeters and Boer (2000) analyze culture and flight safety in military aviation.

Similar to Merritt's work, another study on cultural background and flight deck operations (Hutchins, Holder, & Pérez, 2002), funded by Boeing, emphasizes "that everything a person does is essentially a part of his cultural make-up." The authors note that there is an ethnocentric perspective inherent within mono-cultural and monolingual flight crews. As an example, the authors argue that "an American-born crew flying an American-made airplane in American airspace is a system that is loaded with important cultural properties. But 'seeing' these properties is nearly impossible for someone who has not mastered another language and culture." On the other hand, Mumaw and Holder (2002) acknowledge that cultural differences exist, but are "skeptical of claims of the direct influence of presumed cultural traits on individual pilot behavior in the flight deck."

Much of the research into the effects that culture has on aviation communications is based upon the framework established by Geert Hofstede's (1980) extensive study on cultural dimensions. Hofstede originally classified culture into four dimensions: Power Distance (PDI), Individualism versus Collectivism (IDV), Masculinity versus Femininity (MAS), and Uncertainty Avoidance (UAI). In countries that are ranked high on "power distance," factors such as age and rank contribute strongly to how resource management (CRM) skills influence the decision-making process. Another cultural dimension is uncertainty avoidance (see also Chapter 4). Teams with high UAI values believe that rules and procedures are in place for safety reasons and should be adhered to. For example, a study conducted by Helmreich and Merritt (1996) shows that 19 percent of all respondents from one culture to 85 percent in another culture required written procedures for all in-flight situations.

3.3. Language as a human factor in aviation or LHUFT (Mathews, 2012; Mathews, Carson, & Albritton, 2018)

When aviation safety experts and psychologists began to introduce the importance of human factors to the aviation industry, they were met with some resistance (Helmreich & Merrit, 1996; Helmreich & Foushee, 1993). Some pilots, it is reported, referred to new required CRM courses as "charm school" for pilots. In contrast to the physical and mechanical aspects of flight with which many pilots were more comfortable, human factors and communications were perceived to be "soft" science. Nonetheless, over time, with the support of academic human factors specialists—in particular psychologists who worked with Helmreich in the *Human Factors Research Project* at the University of Texas, who were publishing the results of research studies highlighting not only the role of the human element in aviation safety but also the safety impact of intervention training—the industry has accepted the integration of human factors into models of flight safety (Helmreich, Merrit, & Wilhelm, 1999). The evolution of CRM into more current TEM programs, embedded in the overall SMS of an airline, has been globally accepted, and an entire ICAO Annex 19—SMS—is devoted to supporting the integration of human factors awareness into airline and air navigation service safety systems and procedures.

To some extent, but with a few additional hurdles, the achievement of industry acceptance of the need to focus attention and resources on the study of a range of language-in-aviation safety issues faces some of the same challenges today as the early advocates for a human factors approach to aviation safety faced forty years ago. The most significant challenge to incorporating language awareness and training into aviation safety systems is the comparatively intensive training that is required to achieve a level of English proficiency for effective radiotelephony communications. When airlines were provided evidence of the need for, and effectiveness of, CRM training, they were able to respond with short, one-day training courses: The company also instigated a one-day TEM training course for all its pilots. Trainers introduced the concepts of Threat and Error and then debriefed the LOSA [Line Operations Safety Assessments] findings. As a result, pilots gained a different perspective of safety performance at their airline as reflected in organizational threat and error prevalence and management rates. The pilots responded positively, analyzing the data for reasons and using what they learned to proactively enhance their own performance (Merritt & Klinect, 2006).

The training was shown to be effective, with a positive effect on how pilots responded to managing operational threats and errors, and these outcomes were accomplished with a single-day course. The aviation industry speaks about flight training in terms of hours of training required: achieving a private pilot's license may require sixty hours of flight training. In the United States., a pilot must have 1,500 hours of logged flight time in order to qualify to fly as a copilot on an airline. In other countries, a pilot can qualify as copilot with as few as 300 hours of logged flight time. Recurring annual airline training may be three or four days. Training to transition from flying a propeller engine aircraft to flying aircraft with jet engines requires a few weeks.

These training time frames contrast markedly to those required for language learning which may entail months, depending upon a multitude of factors, including starting levels of language proficiency. Complying with ICAO language standards requires an unprecedented commitment of training time and resources from the industry. In fact, even the term, "training," in particular, "language training," misleadingly suggests, perhaps, that language learning and the acquisition of the necessary level of language proficiency are something that might be accomplished completely within the scope of one or another short course. The misunderstanding of the contrast between the training required to increase human factors awareness, or even the hours required for flight training, and the teaching and learning required to make measurable improvement in English language proficiency is one of the most difficult challenges impeding global compliance with ICAO language standards, even without taking into account the need to identify and teach English that is required for other aviation contexts, such as English reading proficiency for effective flight training in English-as-a-second-language contexts. This misunderstanding is likely due, in large part, to a lack of understanding within the aviation industry of the intricacies of the process of language acquisition, something that industry should not be criticized for any more than linguists should be criticized for an initial absence of knowledge about the technicalities and complexities of the aviation industry. The implication, clearly, is that enlightenment and information sharing in both directions must become part of the collaborative work plan for both the aviation industry and linguists as they cooperate to address the critical needs.

LHUFT: Language as a human factor in aviation or LHUFT (Mathews, 2012; Mathews et al., 2018), then, first seeks a broad and more accurate understanding the role of language in aviation safety, how language, language use, language proficiency, and culture affect aviation safety. As previously emphasized, language is a factor across a very wide range of aviation communication contexts, and includes reading proficiency for pilots, controllers, mechanics, and speaking and listening proficiency in many contexts beyond radiotelephony communications, such as flight and recurrent training. An important positive consequence of the adoption of ICAO LPRs for radiotelephony communications is that it has fostered an increased focus on language in all aspects of aviation communications.

3.3.1. Human factors challenges facing the aviation industry

As discussed, the length of time, effort, and resources that is required to achieve ICAO Operational Level 4 is one challenge. Another challenge specific to the implementation of ICAO language proficiency requirements and other language requirements yet to be identified in the aviation industry is that language is complex, and language teaching, testing, and materials development are distinct professional activities with academic requirements that lie very much outside the domain of traditional aviation expertise. Applied linguists spend two or more years in graduate programs to acquire an understanding of just some part of the vast field of linguistics, in order to be effective language teachers, teacher trainers, curriculum developers, or language testers. The industry cannot draw upon insider expertise to address the need for language teaching and learning programs, for curriculum development, or for testing programs. The industry needs the input of the applied linguistics community.

The corollary to the absence of internal aviation-industry language acquisition, teaching, or testing expertise is that effective aviation English language teaching and testing programs must be informed by an aviation operational awareness and knowledge, which is not a part of the typical linguist's training or knowledge base. Consequently, as suggested above, the synergy needed to accomplish the mutual objective of improved aviation safety will require the two-way, preparatory, mutual "education" of both the aviation industry and language acquisition experts. This is a third challenge facing the aviation industry in efforts to address a globally expanding need for English language proficiency for safe and effective flight training and operations. Aviation English is one of the most technology-dependent fields of any English for Specific Purposes program, and inarguably the field with the highest stakes. Because the stakes are so high in aviation language teaching and testing—with not only safety at stake, but also pilot and controller careers and the economic stability of airlines at stake—aviation English teaching and testing programs, and curriculum and learning materials should represent best practices in language teaching, bringing an academic perspective to the content-based and task-based teaching programs that are most suited to aviation language teaching. Content-based and task-based aviation English teaching (see Part Three) requires developers and teachers who are comfortable incorporating technical and operational topics into their curriculum and classes. It is important that such content be technically accurate; teachers risk losing learners' confidence if the technical content of an aviation English class is inaccurate or lacks realism and relevance. It is not difficult for applied linguists and language teachers to acquire "aviation familiarity" but it does require an investment of time and interest, it is an essential component, and there are still limited well-established "aviation familiarity" teacher training programs; an example of this is from Latitude Aviation English Services (http://www.latitude-aes.aero/). The field of aviation English can only be described as still in development; this is the fourth significant challenge facing the aviation industry's effort, that is, to not only comply with ICAO LPRs but also to provide the English language teaching and learning resources to other segments of the aviation industry to enable that to happen.

3.3.2. Research areas and directions

Many areas of language in aviation remain unexplored. Just as the academic research community supported the industry in the development and implementation of human factors awareness, regulation, and training, the academic applied linguistic community can turn a linguistic spotlight on problems of language in aviation. Research is needed in language as a human factor in aviation that draws on the expertise of both aviation operational and human factors experts working in concert with applied linguists. One area of potential research is an investigation into how language differences and language proficiency impact cross-cultural CRM and/or TEM. With the anticipated growth in the aviation industry that Boeing projects, Asia alone will need twice as many pilots in the next twenty years as it currently has to meet growing demand. For many decades, airlines could operate by hiring pilots from their own ranks of national pilots. In recent years, however, many airlines hire foreign nationals to fly alongside national pilots. Increasingly, cross-cultural and cross-linguistic flight decks are not unusual (Croft 2015; Weigel, 2018). Thus, CRM and TEM training has had a positive impact on the safety and efficiency of flight. Helmreich and Merritt researched how well lessons from CRM training apply cross-culturally (1996). A 1997 study by Merritt and Ratwatte specifically examined how language and cultural diversity affect flight deck communications. The effects not just of cross-cultural communications but also multilingual communications, such as the influence on "Power Distance" when a senior English-as-a-second-language speaking captain interacts with a junior English-as-a-first-language first officer, are areas that merit investigation. There are many such areas of potential linguistic research to explore in aviation.

If flight training programs make formal concession to the challenges that participants with limited English proficiency face, it is not widely known that they do so. Instructors and course developers are often not provided with training to support awareness of the linguistic and cross-cultural challenges that the majority of their trainees face. Instructors and SMEs in aviation training sessions or seminars are not usually trained to automatically incorporate strategies to address the language gap that often exists. This is true in flight training programs in English-speaking locations and in thousands of recurrent or "jet transition" training sessions every year.

These topics represent just a very few that deserve linguistic research. Language proficiency can support aviation safety or its absence can be a threat to aviation safety. Identifying how, where, and under what circumstances language impacts aviation safety and how we can best address linguistic safety gaps when they are identified remains very much to be determined.

3.3.3. Accident investigation

One important reason for more and better linguistic research into aviation communications is to derive practical applications and tools to better inform aviation safety experts about the impact of language in aviation to support and assist accident investigators and other safety advocates in their work. Prior to the advent of Line-Operations Safety Audit (LOSA), most lessons about aviation safety were learned when something went wrong with a flight—a reactive approach. LOSA was initiated as a process to analyze "normal" flights to extract the safety lessons when things go right. Nonetheless, the accident investigation process remains a key part of aviation safety; many lessons can be deduced from an expert, thorough, and well-informed search for all possible factors that led or contributed to an accident.

Accident investigators are pilots, controllers, engineers, and other operational and human factors specialists who are trained to deliberately and systematically examine all aspects of an aviation accident, including possible human factors, in order to gather and analyze evidence. Accident investigation experts go to great expense and effort to recover and, sometimes, reconstruct the parts of the aircraft, in particular the "black box," or Cockpit Voice Recorder. They often test in laboratories or flight simulators theories about possible causes. Because language itself is a complex aspect of human performance and because the potential negative influence of language on aviation safety can be subtle, it can be difficult for aviation operational and human factors experts to notice all but the most obvious of language-related factors in an accident. Compounding the challenge of investigating language with the same degree of systematic and deliberate expert review with which all other human performance factors are investigated, investigators are rarely, if ever, provided training or tools that would support such systematic and expert review. Accident investigators systematically gather and consider information about the physical and psychological states of the pilots or controllers involved in an accident, including blood toxicity, marital and financial status, and sleep patterns in the days preceding the accident. Conspicuously absent from the investigation, from the linguist's perspective, however, is information about language use or language proficiency—something that is not routinely, formally, or expertly addressed in most accident reports.

Accident investigators do not seek to assign blame; the findings and conclusions of accident investigations are based on the facts and evidence uncovered during the investigations. The findings are important because it is through accident analysis that causal, contributory, and latent factors and other safety deficiencies are identified and made public. Recommendations from accident investigation boards, such as the National Transportation Safety Board, prompt industry response. Dumitru and Boşcoianu (2015) point out that, in the modern approach to aviation safety, human error is not the end but rather the starting point in investigating and preventing aviation events and accidents. If accident investigators miss the more subtle or less visible—but nonetheless critical or contributory—role of language in an accident, they have missed an opportunity for the industry to better understand the importance of language and language proficiency to aviation safety. If evidence of language factors is sensed or suspected but not pursued, or clearly and conclusively identified, but not presented at the level of a "Finding" or identified as a contributory or latent factor in the chain of events that led to the accident, then the industry misses an opportunity to have highlighted, in more specific terms than has been the case thus far, the role of language as a possible contributory or latent factor in accidents and/or adverse events.

3.3.4. *Taxonomy of communication and language as a human factor in aviation*

Accident investigators should be provided with more tools and training support to help them understand and analyze language factors, including an understanding of the more subtle ways that language impacts safety. The Human Factors Analysis and Classification System (HFACS) is a widely used human factors aviation taxonomy. The HFACS taxonomy lists not only "Unsafe Acts" but also "Preconditions for Unsafe Acts" (Shappel & Wiegmann, 1996). *Failure to communicate/coordinate* is listed under CRM as an Unsafe Aircrew Condition. A linguistic perspective will take this as the starting point from which to launch additional research and analysis in order to determine *why* communication and/or coordination failed. Although no taxonomy may fully account for the full range of language in aviation factors, a Taxonomy can be one useful tool for accident investigators, safety advocates, as well as language teaching program developers and managers. A *Taxonomy of Communications in Aviation* has been developed by researchers at Embry-Riddle Aeronautical University as an aid to support accident investigators and safety experts, and researchers interested in language use in aviation.

Based on a draft Taxonomy developed in 2012 by Mathews during the early stages of a review of an aviation accident database, the Taxonomy takes "Communications" as the "starting point" of inquiry and analysis, identifies four major categories of communication factors in aviation: technical, procedural, cultural, and linguistic. The Taxonomy is designed as if for the layperson's use, accurate but simple enough to facilitate use by aviation operational specialists with no specialized linguistic expertise. Each communication factor type is sub-divided into categories or examples of communication problems or errors in that category. Because the aviation industry has long addressed cultural factors in aviation, cultural issues that seem most related to language and communication are highlighted. The taxonomy is not exhaustive, but is intended to be representative of major communication factors as identified in the Embry-Riddle accidents and serious incident review (a project in preparation by Mathews and her research collaborators). The taxonomy is drawn from a linguistic review of accidents in which English proficiency may have contributed to the chain of events that resulted ultimately in the accident.

A brief review of the major categories of communication factors outlined in the Taxonomy (Table 3.1) as shown below highlights a few of accidents that can be attributable, in part to communication factors.

Table 3.1 Proposed Taxonomy of Communications in Aviation adapted for this book from Mathews (2012) and Mathews et al. (2018)

Taxonomy of Communication Factors in Aviation

1 **Technical Factors** that impede communication: equipment failures; or static or noise on the radio frequency.

   Example: a loss of the transponder signal Midair Collision, 2007

2 **Procedural Factors** that impede communication: a blocked transmissions; omitted call signs; or the use of two languages in a single operating environment.

   Example: Use of both French and English at a French airport, Runway Collision, 2002

   Example: The runway collision at Tenerife, Spain. The communication from Pan Am to confirm he was still on the active runway was "blocked" by another speaker attempting to communicate at the same time.

(Continued)

3. **Language Factors** are sub-divided into ***Spoken Language Factors*** and ***Written Language Factors*** that may impede communication. Spoken Language Factors include Speaking and Listening. Written language factors include Reading proficiency and Writing.

SPEAKING

Pronunciation—A speaker's pronunciation may interfere with comprehension of a message or make a message unintelligible.

Example: Helios, 2007; ground staff reported inability to comprehend the Captain's message, possibly due to pronunciation issues.

Structure—Errors in structure (or grammar) may interfere with an intended message.
Example: "We just running out of fuel": Avianca First Officer communicating his aircraft's emergency fuel situation to ATC, New York, 1999.

Vocabulary—Vocabulary issues may interfere with an intended message.

Example: Helios Captain asked about the circuit-breaker when the aural alert was related to the a problem in pressurization, 2007.

Fluency—Spoken language which is not fluent may interfere with transmission of a message, or a lack of Fluency may possibly indicate English language proficiency below ICAO Operational Level 4, and should be investigated as a possible language factor.

Example: The publicly available audio recording of the pilot in a general aviation accident in Florida exhibited speech that was not fluent as she attempted to communicate with ATC during a flight in poor weather. While the accident was not attributed to language use, the NTSB urges that safety deficiencies, even unrelated to an accident, be identified and included in the Findings of an accident investigation report.

Failure to communicate—Sometimes there is a simple failure to communicate which may indicate inadequate overall language proficiency or a failure in comprehension; a failure to communicate should be investigated as a possible language factor.

Example: Two air traffic controllers along an air route in Brazil failed to communicate critical and required information regarding an aircraft's loss of a transponder signal: the loss of the transponder was determined to be a causal factor in the subsequent midair collision. The failure to communicate the transponder failure was noted, and a number of possible reasons for the communication failure explored; however, the there is no information in the report to indicate that language proficiency was considered as a possible reason for the communication failure. 2006.

LISTENING

Listening—Listening proficiency permits comprehension of a message. "Radio silence," or no response to a message, that is a "*Failure to Communicate*," may also be indicative of a listening comprehension problem. Repeated requests to "Say again" may indicate a comprehension problem.

Inaccurate Read-back, or a response which does not align to a speaker's message, may be indicative of a listening comprehension problem or it may indicate inadequate language proficiency impeding a listener's ability to respond.

Example: An Air China First Officer responded to a controller's inquiry, "Can you landing?" with an answer that was not related to the question. The controller twice queried the pilot, and twice the pilot responded but his response did not align to the question asked.

Written Language, or Text
Reading
Manuals—pilots, air traffic controllers, and airplane mechanics and engineers rely on much written material, on board, during flight; on the ground, in pre-flight planning; on ground for maintenance repairs and checks; during initial, jet transition, and annual recurrent training, inter alia. It is more difficult to uncover possible reading problems related to aviation safety because they are rarely recorded, in contrast to radiotelephony communications, captured by Cockpit Voice Recorders. Nonetheless, reading proficiency can interfere with comprehension of important safety information.

Possible example: Turkish Airline in Paris, 1970s
Value Jet accident, canisters, 1996.

Writing
How text is written impacts aviation safety. Manuals, datalink, graphic presentation of information

Chinese Accident Investigation Report noted that some of the "repair assessment program" content was "vague to non-English speaking persons." ("China Airlines Accident Report," 2002, p. 95).

4 Cultural Factors
The Taxonomy also seeks to account for cultural factors affecting communication. Human factors in aviation has drawn very much from the research of Hofstede's (1980) Cultural Dimensions to understand cultural issues in CRM and TEM. The 2018 Taxonomy seeks to highlight cultural factors that most directly affect language and communications.

- Sociolinguistic Factors, including issues of language apprehension in aviation communications.

- National Culture, including issues of power distance.

- Organizational Culture, most particularly the safety culture and safety training programs of an organization.

- Individual Culture effects include a lack of situational awareness regarding language or cultural differences, and personal factors affecting communication.

A Taxonomy of Communication Factors in Aviation can be seen a starting point to support better industry understanding and analysis of possible language issues in aviation. More linguistic research into language issues in aviation will undoubtedly help refine, improve, and increase the tools we can provide to assist the aviation industry, in addition to the Taxonomy.

3.3.5. Aviation English testing

The 2003 introduction of strengthened ICAO LPRs, with a 2008 deadline for full compliance that was, for the most part, missed by large sectors of the industry, imposes much more rigorous testing and training requirements on the industry (ICAO AS26-11 and AS27-10). Despite the struggles to implement ICAO LPRs that ICAO documents in Assembly Resolutions A26-11 and A27-10, much progress has been made, and ICAO language Standards are the global policy that will continue to drive English language testing and teaching programs in aviation. Without ICAO language Standards and Recommended Practices, the industry could not invest in, or support the development of, aviation English training and testing programs.

Compliance: The current focus of the industry is, understandably, on compliance with ICAO Language Standards and Recommended Practices, and there are many challenges facing the industry. Although many aviation English tests have been developed, aviation language testing is largely unregulated. Civil Aviation Authorities are responsible to certify language testing programs for national use, but they rarely have access to support and expertise from language testing specialists to aid in the evaluation of these instruments. What is needed now in the aviation industry is the full breadth of language teaching, teacher training, and testing support programs that are available in the English for Academic Purposes context. No one who understood second language acquisition expected that it would be simple, easy, or quick for the industry to achieve global compliance with ICAO LPRs. However, without ICAO LPRs, there is no reason to expect that the industry would have made even the progress that has been made.

3.4. Conclusion

Language use and language proficiency affect aviation communications in many, varied, and profound ways. Language is inevitably and inextricably a fundamental component of aviation communications. If effective communication is a skill that can be learned, developed, and improved (Flin, O'Connor, & Crichton, 2008), it is only possible when the individual has a minimum operational language proficiency. The benefit of CRM, although difficult to quantify, is universally acknowledged; no responsible aviation professional will dispute the importance of CRM training and awareness. The same will be true with respect to language issues in aviation. There are some initial resistance and widespread misunderstanding. With time—and with increased input from the applied linguists and language teaching specialists—as well as with better identification of language factors in aviation communications, will come greater acceptance of the responsibility of all aviation operational and safety professionals to properly and more fully understand and utilize knowledge about the ways in which language affects all aspects of aviation safety.

Within the otherwise tightly regulated aviation industry, the gaps around language issues are an anomaly. Full-motion simulators and rigorous, well-regulated training requirements ensure that pilots are among the best trained of any group of professionals. But there are safety gaps in the industry related to language issues that would not be tolerated in other parts of aviation training or regulation. Dumitru and Boşcoianu (2015) point out that the human element is the most flexible, adaptable, and valuable component of the aeronautical environment, but, at the same time, ironically, flexibility and adaptability inherently include variability which—unless managed effectively—can result in vulnerability. In an era in which safety experts acknowledge that there is already an overall safety-oriented aviation industry, as far as equipment and personnel performance are concerned, there does remain at least one realm within which additional gains may be realized. An industry-wide commitment to developing academically well-informed, aviation-specific, content-based, English language teaching and learning materials dealing with topics of compelling relevance and interest to learners remains as perhaps the single most productive and cost-effective strategy the industry can take to move the safety needle even higher. Increasing the safety awareness and language proficiency of pilots, controllers, and airline mechanics via content-based language teaching modules dealing with human factors and safety topics of high interest will provide aviation professionals the best means for increasing English language proficiency. Transport Canada's *Human Factors for Aviation: Advanced Handbook* points out that, " … at the heart of CRM is communication." It is the important role of the applied linguist in aviation to help the industry understand that at the heart of communication is language.

Chapter Co-Authors

Aline Pacheco is an Associate Professor at the Pontifical Catholic University of Rio Grande do Sul (PUCRS), in Brazil. She holds a PhD in Linguistic Theory and Analysis from the Federal University of Rio Grande do Sul (UFRGS), a master's degree in Applied Linguistics and a bachelor's degree in Letters from PUCRS. She has been working as an English teacher since 1993. She is currently teaching at the School of Aeronautical Science. Her major interests are English for Aviation and English as a Foreign Language.

Angela C. Albritton is the Director of Military Relations and Strategic Initiatives for Embry-Riddle Aeronautical University's Worldwide Campus. She focuses on advancing and supporting the university's strategic goals and special projects by identifying, evaluating, and championing opportunities for external collaboration and growth. She holds a master's degree in Aeronautics from Embry-Riddle with emphasis on Human Factors and a bachelor's degree in Management from the University of Maryland and a University College and a teacher certification program from the University of New Mexico.

PART TWO

Research

4

English in Global Aviation: Research Perspectives

with Justin Taylor and Rachelle Udell

4.1. Introduction

Research into the language of aviation, especially as it has shifted increasingly to English-based communications, is still relatively limited—surprising, and even concerning, perhaps, considering the size and global scope of the industry. There are, however, encouraging developments over the past twenty years, with many linguists and English instructors collaborating with aviation subject matter experts and training institutions and universities around the world to collect data and analyze needs and ways of effectively meeting these needs. The results of this collaborative work are being shared in publications as well as at professional and academic conferences. Major academic conferences today, such as the American Association for Applied Linguistics (AAAL) Conference, held every year in the United States (or Canada) or the TESOL Convention also in the United States, have included papers on aviation communication. More specialized professional meetings and colloquia such as the Language Testing Research Colloquium (LTRC) within the International Language Testing Association (ILTA) or the American Association for Corpus Linguistics (AACL) Conference have also included papers on assessment, evaluation, and discourse-analytic studies of the language of aviation. LTRC 2010 featured a colloquium on "Testing Aviation English," and AAAL 2012 included a well-attended panel on "Language Tests and Language Policy: The Case of Aviation English."

4.1.1. International Civil Aviation English Association

The International Civil Aviation English Association (ICAEA) is the primary international academic and professional organization dedicated to the study of aviation English and its testing and pedagogical applications. The ICAEA has been working to provide resources and tools that can be of use to aviation English trainers and examiners around the world. The ICAEA was organized by Fiona Robertson, with her colleague and friend, Philip Shawcross, to host a series of Aviation English Forums in Paris in the 1980s. The success of those forums led to the creation of ICAEA as a not-for-profit association in 1991. Dedicated to improving aviation safety, Robertson and then Shawcross led ICAEA in its primary mission to organize and host international workshops and conferences in support of aviation English language teachers and, later, testers. For more than thirty years, ICAEA has been run by a small group of dedicated volunteers, and was granted a nongovernmental organization (NGO) status by the United Nations in 2011. ICAEA has collaborated with and supported ICAO on a number of aviation English projects. Under the current leadership of the third ICAEA President, Michael Kay, ICAEA is developing a series of global ICAO LPR Test Design Workshops (see official website at: https://www.icaea.aero).

Three of ICAEA's most notable contributions include (1) the 2009 ICAO Circular 323, Guidelines for Aviation English Training Programs, (2) the 2012 ICAO Rated Speech Sample Training Aid (RSSTA), and (3) the recent ICAO LPR Test Design Guidelines. All of these tools were developed to assist CAAs and organizations involved in the testing of ICAO LPRs. ICAO Circular 323 continues to provide valuable guidance to both emerging and existing training programs for aviation English, exploring four key areas: (1) training design and development, (2) training delivery, (3) instructor profiles and backgrounds, and (4) instructor training.

The RSSTA is a free online resource that has helped those around the world tasked with evaluating aviation language proficiency, and is mentioned as a necessary training tool in many regulatory documents advising the implementation of the LPRs in operational environments. (See FAA AC 60-28B, Transport Canada AC 401–009.) Finally, the Test Design Guidelines are a recent response to the predictable differences in testing standards and practices around the world that have arisen since the early 2000s when ICAO published information on the implementation of the LPRs, available in ICAO Document 9835. Although ICAO LPRs themselves are generally globally accepted, the quality of tests is largely inconsistent (Mathews, 2014). The Test Design Guidelines contain eight criteria responding to the unique challenges of aviation language testing which are currently detracting from the utility and effectiveness of ICAO LPRs and, consequently, from aviation safety. The guidelines are delivered to stakeholders through a series of practical, regional workshops by providing a framework which CAAs can use when selecting an LPR test and for developers to use when designing the test instruments themselves. ICAEA responds to areas of need in the world of language in aviation safety and provides timely and constructive resources to entities, such as airlines, air navigation service providers (ANSPs), or flight schools.

In addition, ICAEA hosts international conferences, workshops, and forums around the world in an effort to bring together the small community of aviation language professionals to share ideas, present research, and plan for the future of aviation safety. Conference themes change yearly, with the three most recent conferences focusing on meeting the training needs of pilots and controllers, exploring ICAO LPRs on their tenth anniversary of adoption, and taking a look communication as a human factor in aviation. In 2016, the conference's focus on aviation English training was supported through interactive workshops which encouraged engagement and interaction of participants, examining issues such as learner motivation, curriculum content and delivery considerations, and trainer competencies. Focusing on training, the 2017 event explored issues surrounding the LPRs, including test washback, attitudes toward testing and learning, and the overall effects of the LPR implementation. Hosted by Embry-Riddle Aeronautical University, the most recent 2018 conference took a more research-based approach, and explored issues beyond the LPRs, including communication strategies and the effects of language and culture in aviation safety. ICAEA's 2019 conference will explore new territory, including the training needs of those not formally covered by ICAO LPRs such as *ab-initio* pilots and air traffic controllers (ATCs), as well as ground staff and aircraft maintenance personnel. Attended by participants from over thirty countries, these events bring together hundreds of individuals from airlines, flight training organizations (FTOs), ANSPs, civil aviation authorities (CAAs), universities, and testing and training providers to learn about new trends and research in the field of aviation language.

Because of the physical separation of most of ICAEA's members, several virtual platforms exist to allow members to distribute and discuss information related to English in aviation. In 2017, ICAEA's first Research Group (ICAEA-RG) was formed to provide a forum for collaboration and conversation among those currently involved in academic research in the aviation context in areas such as ESP, Corpus Linguistics, English as a *Lingua Franca*, and Intercultural Communication. Members of the ICAEA-RG have conducted and published research on aviation language testing, pronunciation features of aviation English, and the compilation and analysis of corpora in aeronautical registers. Finally, ICAEA encourages discussion on issues of current interest through the ICAEA Forum on LinkedIn. This active forum allows all members to remain connected and learn from fellow aviation language professionals.

4.2. Pilot-ATC interaction research

In 1994, Steven Cushing published *Fatal Words: Communication Clashes and Aircraft Crashes*, which focused on an analysis of interactions and miscommunications primarily between pilots and ATCs. Cushing's monograph highlighted the March 27, 1977, KLM and Pan Am 747 collision on a crowded, foggy runway in Tenerife, the Canary Islands, resulting in 583 casualties (also mentioned in the previous chapter). He explored main causes of this accident, especially the documented miscommunication between the pilot and ATC; for example, as identified by Cushing, "We are now at takeoff," radioed by the pilot, meaning that the plane was lifting off, but misunderstood by the controller to mean that the plane was *waiting on the runway* for takeoff clearance.

In Cushing's view, miscommunication in general, and especially with pilots and ATCs who do not share the same first language backgrounds, has led to dozens of aircraft disasters, and he proposes solutions for preventing them. He further examines what he considered as ambiguities in language when aviation jargon and colloquial English are mixed, when a word is used that has different meanings, and when different words are used that sound alike. Examples of some of these items, grouped according to language features such as phonology, vocabulary and grammar, and speech intonation, are provided in Table 4.1. To address these potentially fatal communication problems, Cushing proposes a visual communication system and a computerized voice mechanism to help clear up confusing language in aviation interactions.

Table 4.1 Sample analysis of English in aviation communications from Cushing (1994)

Language Feature	Speakers	Sample Communicative Issue
Phonology and understanding of pronunciation	ATC to another ATC	[ATC 1] I'll let you know. (understood as) Let him go [by ATC 2]
	Instructor to student pilot	[Instructor] Last of the power (i.e., reduce power) (understood as) Blast of power (i.e., **increase** power)
Vocabulary and grammar	ATC to pilot	[ATC] Descend two four zero zero (i.e., 2,400 feet) [Pilot] OK. four zero zero (i.e., **400** feet)
	Copilot to pilot	[Copilot] Cleared to seven (i.e., descend to 7,000 ft) [Pilot] Cleared 27 (i.e., permission to land on RW 27)

Language Feature	Speakers	Sample Communicative Issue
Speech: Intonation (sample focus)	ATC to pilot	[ATC] Traffic, level at 6,000 (i.e. descend to 6,000 ft) [ATC] Traffic level at 6,000 (i.e., you've got another aircraft at 6,000 ft)
	Instructor to student pilot	[Instructor] Back—on the power. (i.e., reduce power) [Student pilot] Back on the power. (i.e., increase power)

Following Cushing's work, more detailed analyses of pilot-ATC discourse have been published, focusing on a variety of approaches and target linguistic areas. For example, Barshi (1997) and Barshi and Healy (1998) explored the effects of ATCs' pausing and speech rate and use of related suprasegmental features on pilot responses and comprehension. They examined, in particular, readback accuracy, and related studies have included non-native English-speaking pilots and other contexts of aviation communication interaction. More than a decade later, Kim and Elder (2009) conducted a detailed analysis of communication between a native English-speaking pilot and a Korean ATC, revealing that native speakers need to develop effective speech accommodation skills in communicating with non-native speakers in this communicative domain. Featured in the study was a four-minute and thirty-second communication between an American, native English-speaking pilot and a native Korean-speaking ATC. Because of adverse weather conditions, the pilot was low on fuel and was requesting to land at Shanghai airport to refuel before continuing on to Hong Kong. While in Korean airspace, the pilot requested a change to come into Osaka. Analysis of the recorded transmission revealed that nearly all of the pilot's transmissions utilized plain language rather than standard phraseology, which caused confusion for the Korean ATCs. Kim and Elder noted that communication in the aviation context is a complex matter and responsibilities for its success (or failure) are commonly shared among all participants involved in the communication, regardless of their language background. In other words, communicating effectively in aviation contexts is more than a simple matter of using standard phraseology. Other language resources and "awarenesses" must be employed when needs arise (Kim, 2011).

Bieswanger (2016) examined whether or not standard phraseology and plain language are two separate registers of spoken aviation English based upon the parameters for register analysis detailed in Biber and Conrad (2009). Standard phraseology, as the term suggests, was found to be restricted to standard situations likely more common to a variety of locations and circumstances, in contrast to more variable plain language, and that pilots and ATCs must *code switch* frequently between the two registers. A study by Bowles (2014) described the organization of ATC communication from a qualitative conversational analysis of seventeen exchanges between ATCs who did not share the same first language. The focus here was to explore how ATCs negotiate their understanding of operations with other controllers. Bowles noted evidence of atypicality from a procedural perspective which may contribute to potential miscommunication on the job with international pilots. For related studies of aviation language assessment, Farris and Turner (2015) provide an extensive summary of research in aviation communication highlighting analytic approaches that have been commonly employed, as well as future directions in exploring the evaluation of language performance by pilots and ATCs beyond ICAO LPRs.

Another major, recent publication is Dominique Estival, Candace Farris, and Brett Molesworth's (2016) monograph, *Aviation English: A Lingua Franca for Pilots and Air-Traffic Controllers*, which investigates ATC-pilot communications from a performance perspective based on several socio-cognitive factors, (mis)communication challenges based on the merging of human and equipment contexts, and a description of Aviation English as a language variety. They also discuss how speakers may deviate from mandated phraseology from (or due to) contextual factors in actual practice. Estival, Farris, and Molesworth have collaborated on a number of related studies previously or have conducted their own extensive research, primarily based in Australia. For example, Molesworth and Estival (2015) investigated the relationship between four flight conditions, including (1) pauses between items in ATC communication, (2) the number of items in a transmission, (3) pilot workload, and (4) the level of radio congestion and English L2 pilots' ability to communicate effectively without errors. They found that radiotelephony (RTF) communication is both complex and "cognitively taxing," especially for English L2 speakers who face increased pressure when language performance is combined with other stressors related to high workload. When extremes in flight conditions are present, that is, few to no pauses between items, high number of items in a transmission, high workload, and high levels of radio congestion, etc., the language performance of all pilots—NS and NNES alike—deteriorates. Molesworth and Estival caution that the assumption that a pilot's communication skills will develop as he or she accumulates hours of flight experience is an erroneous and dangerous one. They suggest that communication performance should be judged upon a pilot's recall license qualifications instead.

In most of these pilot-ATC studies, generalizations and conclusions include discussions of the effects of training, especially for international pilots and their language learning background in English, relative to their ability to follow ICAO standards and requirements. Contextual factors impacting aviation communication are highlighted, especially noise and various other conditions that contribute to stress, fatigue, and (mis)communication, that is, human factors. Studies of non-native English speaker (NNES) vs. native speaker (NS) interactions in aviation have also incorporated related focus on flight conditions on the communication accuracy of pilots with different first language backgrounds and with varying flight training levels. Relevant results have shown that increased pilot workload adversely affected communication performance, including that of NSs; increased ATC speech rate was problematic for NNES pilots with low pilot qualifications; and including four or more content items in radio transmission may adversely affect communication accuracy even for NS and those with higher flight qualifications (Jang, Molesworth, Burgess, & Estival, 2014; Wu, Molesworth, & Estival, 2018).

And finally, Anna P. Borowska's (2017) *Avialinguistics: The Study of Language for Aviation Purposes* describes what she conceptualizes with the term *avialinguistics*—an interdisciplinary branch of applied linguistics recognizing the dynamic role of language (especially English) for aviation discursive practices. Following a detailed explanation of the history of Aviation English, she suggests that the term *avialinguistics* be adopted to describe "the study of aviation language in all its professional aspects in relation to practical problems" (p. 55). This includes pure linguistic studies of the characteristics inherent in aviation language itself as well as analyses of aviation language dynamics as it is actually utilized by professionals in the field; how aviation language is taught, tested, and rated; and how training materials and other language teaching methodologies are prepared and employed.

Borowska explored a range of data and linguistic markers from syntactic structures and standard lexemes of phraseology, the linguistic nature of Plain Aeronautical English (PAE), and related markers beyond the prescribed system of language, for example, code-switching, expectation bias, or insufficient command of general English by NNES. She also includes a study of pilot-ATC communication designed "to elicit [self-]assessments of radio communication experience in Aeronautical English" from participants via survey analysis. Borowska's approach is to describe in-depth current international aviation discourse, especially what she considers as its "linguistic errors," and then to provide suggestions for improvement in aviation English communication.

4.2.1. Aviation communication in popular media

In Malcolm Gladwells' (2008) best-selling book, *Outliers: The Story of Success*, he dedicated an entire chapter entitled "The Ethnic Theory of Plane Crashes: Captain, the weather radar has helped us a lot" to document the role of culture and intercultural influence on aviation communication, specifically between pilots and ATCs. Gladwell's primary focus in *Outliers* is to underscore intangibles that result in a select group of people attaining extraordinary success. His thesis is that enormously successful people (or groups of people) do not achieve success and major accomplishments through intelligence and ambition alone. Factors such as work habits;

family, class, and cultural backgrounds; and "being at the right place at the right time" are equally contributory in making extremely positive things come to fruition for individuals. Gladwell argues, for example, that your birthplace can influence whether or not you become a successful international pilot or that the year in which you were born might influence your chances of becoming a Silicon Valley billionaire.

He highlighted Korean Air's KAL 801 crash and multiple other crashes in the 1980s and 1990s that prompted the US National Transportation Safety Board (NTSB) to issue an addendum to safety announcements and company profiles listing all the airline's recent crashes.

He presented a discourse analytic discussion of intercultural communication (e.g., the copilot's utterance, "Captain, the weather radar has helped us a lot," to illustrate indirect speech potentially influenced by role-relations and hierarchy), supported by research in intercultural communication attributed to Dutch social psychologist Geert Hofstede (as briefly referenced in Chapter 3). Hofstede's studies of cultural dimensions' influence on communication are very popular in multicultural business settings and workplace relations.

In April 1999 both Delta and Air France suspended their partnership with Korean Air, and the US Army, which maintains a major troop presence in South Korea "forbade its personnel from flying with the airline." (p. 181). Gladwell asked, "How did Korean Air turn itself around?" Korean Air's safety record has improved dramatically and in 2006, as noted by Gladwell, the company was given the Phoenix Award by Air Transport World acknowledging its major transformation. In 2000, Korean Air brought David Greenberg from Delta Airlines to run flight operations and subsequently evaluate the English language skills of flight crews. Greenberg "set up a program to assist and improve the proficiency of Aviation English" of Korean crews and he also brought in a subsidiary of Boeing called Alteon to take over Korean Air's training and instructional programs—all conducted exclusively in English. The new language of Korean Air was English (p. 218). Greenberg's rationale was that "English was the language of the aviation world" (p. 218). Greenberg wanted to give his pilots an alternate identity; their problem, he believed, was that they were trapped in roles dictated by the heavy weight of their country's cultural legacy.

Not considered or factored in in this argument is the *prescription* that native English-speaking ICAO policy makers have initiated and the potential negative attitudes such linguistic prescriptions (and proscriptions) may have created in other settings that may not correspond or benefit from the Korean model. According to Gladwell, Greenberg knew that cultural legacies matter— that they are "powerful and pervasive and that they persist, long after their original usefulness has passed. But he didn't assume that legacies are an indelible part of who people are. He believed that if the Koreans were honest about where they came from and were willing to confront those aspects of their heritage that did not suit the aviation world, they could change—an opportunity to transform their relationship with their work" (p. 219).

4.3. Global research in aviation English

In the UK and Europe, Barbara Clark's (2017) work, *Aviation English Research Project: Data Analysis, Findings, and Best Practice Recommendations* is an independent research report commissioned by Civil Aviation Authority, Gatwick Airport South, West Sussex. (This report can be easily accessed online, using the report's full title as search terms or at www.caa.co.uk.) Clark investigated pilot-ATC communication issues identified in Mandatory Occurrence Reports (MORs). She then proposed "best practices" to reduce miscommunication resulting in whole or in part by what she identifies as substandard ICAO language proficiency. The crucial interactional problems identified by Clark are: (1) readback-hearback errors by both UK and non-UK pilots and controllers; (2) call sign confusion; (3) language proficiency below ICAO minimum standard (non-UK pilots and controllers); (4) situational awareness reduced due to multilingual radiotelephony (i.e., multiple languages used and heard on the radio); (5) non-standard phraseology use by both NSs and NNESs; (6) potential evidence showing cheating on aviation English exams; (7) grounds to suspect that some NNESs are not being tested but, instead, are granted ICAO Level 4 certificates based on "sweetheart" deals (handshakes, by friends, etc.); (8) ICAO levels of language proficiency, especially Level 4, not sufficiently robust to ensure appropriately clear pilot/controller communication; and (9) poor MOR/language-related reporting culture and underreporting of language proficiency issues by UK pilots and controllers.

Major recommendations proposed by Clark include:

- Increased emphasis in the UK on the importance of reporting language-related miscommunication issues to airlines, the CAA, and CHIRP. Stress that language-related miscommunication issues are as important to aviation safety as any other issue (e.g., mechanical, turbulence, disruptive passengers, etc.);
- Work with ICAO Member States to agree that English has become the language of aviation used in all radiotelephony communications when there is a reasonable expectation that it might be of safety benefit to international traffic irrespective of country or local language;
- Work with European national aviation authorities to reduce language-related miscommunication between UK pilots and controllers based in continental Europe;
- Increase language testing spot checks and expand SAFAs to include language assessment, to ensure non-UK pilots' levels of

English proficiency actually match what their ICAO certificates of proficiency state;

- ICAO language proficiency levels need revising or improving. Current ICAO Level 4 allows for some level of misunderstanding; the evidence is that this safety risk should be managed more effectively. There should be no room for lack of language proficiency in international aviation.

Related doctoral dissertation work: One of the seminal dissertation works on aviation communication was from Barshi's (1997) study of the effects of linguistic properties and message length on misunderstandings in aviation communication from the University of Colorado. Barshi focused on, among others, simulated messages and authentic controller—pilot communications as they corresponded to the order of controller instructions for heading, altitude, and radio frequency (but contained no aviation jargon). Aviation communication with interlocutors from Greece was extensively examined by Theodoros Katerinakis (2014) in his dissertation from Drexel University entitled, *Aviate, Navigate, Communicate: Silence, Voice and Situation Awareness in Aviation Safety*. (See also the annotated bibliography in this chapter for related information.) Katerinakis' main goal was to show how important human communication is "in order to accomplish a safe flight" (p. 12). He conceptualizes a flight as an act of conversation that combines culture with communication of co-present and cooperative actors, in the highly mediated environment of the flight deck. Utilizing this grounded-theory approach, in-depth interviews with aviation informants were conducted and analyzed using discourse analysis and closed reading. The themes that were identified include the role of silence in multiple dimensions as part of an interaction; hesitations and non-verbal and verbal attributes in flight deck speech; and situation awareness across local, transitory, and global communication models. Katerinakis also explored the role of interlocutors' first language as a non-conflictual but complementary communication tool that may facilitate linguistic security (instead of competing with the topical standardized English in aviation). Another completed dissertation on international aviation communication is Moonyoung Park's (2015) *Development and Validation of Virtual Interactive Tasks for an Aviation English Assessment* from Iowa State University. (See also Chapter 5.) This project focused on Korean military ATCs, the assessment of their required aviation English knowledge, skills, and target tasks. Park concluded that the use of immersive interface and simulated target language-use situations in virtual environments may provide military ATC trainees with more authentic opportunities to perform on-the-job the target tasks.

Based in the Philippines, Ferrer, Empinado, Calico, and Floro (2017) conducted an analysis of CVR/QAR transcripts of pilot-ATC RTF communication from the country's Civil Aviation Authority and also conducted interviews with Filipino pilots and ATCs. The researchers found that the lexical items *hold short, go ahead, affirm*, and *priority* have both standard and non-standard definitions in Philippine English, and, if confused during interactions, could result in serious safety consequences. Both pilots and ATC participants in the study indicated that these items are used in non-standard phraseology most frequently in Route Clearance situations. The study concluded that trainees in the Philippines should be instructed on the importance of always using standard phraseology in all RTF communications involving these lexical items. In Brazil, in response to ICAO language requirements for international pilots, the Brazilian Civil Aviation Agency, *Agência Nacional de Aviação Civil* (ANAC), requires that Brazilian pilots should take the Santos Dumont English Assessment test (SDEA) which essentially follows ICAO guidelines. An oral interview for Brazilian pilots is provided in parts such as pilot's background, "interacting as a pilot," "emergency situations," and aviation topics. Brazil, especially the city of São Paulo, is the location of the region's busiest airport, a hub for flights to and from many international locations. The headquarters of TAM Airlines, the largest Brazilian airline, and a major plane manufacturing plant are also instrumental in attracting Brazilians from all over the country who are interested in finding a position in the aviation industry. There are several studies focusing on assessment and training programs in Brazil and indicating a need for a more consistent implementation of English proficiency tests (Machado, 2016). For Brazilian ATCs, another proficiency examination, the EPLIS (*Exame de Proficiência em Língua Inglesa*), from SISCEA (*Sistema de Controle do Espaço Aéreo Brasileiro*), the Brazilian Airspace Control System is required.

4.3.1. Exploring China's civil aviation industry

In 1987, the Chinese government divided the airline operations of the "General Administration of Civil Aviation of China" into a number of state-owned enterprises. A specific regulatory role was retained by what has become the Civil Aviation Administration of China (CAAC) that has since been tasked to focus on a range of language training and communication research areas, especially for Chinese pilots (Dougan, 2016). In 2017, there were 55,765 licensed pilots in China, an increase of 5,261 from 2016 (CAAC, 2018), maintaining a rate of over 10 percent increase per year in the total number of licensed pilots in China from 2013. The country's aviation safety record has also been outstanding in the past decade; from August 25, 2010, to December 2016, the civil aviation industry operated accident-free for 76 months, with an "accumulated 46.23 million hours of safe flight" (CAAC, 2017). Projections show that China will be the world's second largest aviation market after the United States by 2034, with its aircraft manufacturing industry potentially leading the way. In May 2017, the first large Chinese-made passenger jet C919 completed its maiden flight, a key milestone for the country (Ostrower, 2017).

As expected, there has been a major national demand for Chinese pilots, especially captains, over the course of the past ten years, corresponding to the exponential growth of the industry. Chinese aviation universities and flight training institutions have established relationships/partnerships with airline companies and universities in the United States and Europe in jointly developing pilot training programs. From 2013 to 2017, the number of registered student pilots trained in flight training academies domestically and internationally had increased from 3,940 to 5,053 (CAAC, 2018). There are currently nine academic institutions in China that are eligible under the regulations of the CAAC (in particular, CCAR-141) to train and license pilots. One major institution is the Civil Aviation Flight University of China, founded in 1956, which has trained over 10,000 pilots for the Chinese civil aviation industry as of 2018.

Industry regulators and domestic and international airlines in China have been very concerned about Chinese pilots' English proficiency and level of fluency. Chinese pilots themselves, especially those who fly internationally, are equally concerned. According to a survey conducted by Huang and He (2007), 76 percent of the pilots they interviewed considered their English proficiency level as only "average"; 17 percent rated themselves as "below average"; and only 7 percent reported as possessing "above average to fluent" English language proficiency levels. In other English skill areas, 46 percent consider themselves "good" at reading; 30 percent as "good" in oral English; and only 18 percent at "good" in listening skills. Over 68 percent of these pilots wished to improve their oral proficiency in English, while 64 percent were interested in receiving training to improve their listening comprehension. In 2013, a study by Li, Xiao, and Ren further explored English proficiency levels and flight performance evaluations of student pilots from the Civil Aviation Flight University of China by means of questionnaires, training observations, and student-pilot interviews. They reported generally positive results and improvements in language performance after training. From May 2011 to July 2012, after three to six weeks of instruction (from 48 to 128 total contact hours), 33.9 percent of the students passed a similar test to ICAO level 4 when they were evaluated for the first time. There was an increase to 54.6 percent of the students passing the test during the second and third attempts. For student pilots who did not pass language and flight skills training, the study reported that areas for improved and focused support included listening and accurately responding to interlocutors' questions; oral skills, especially in pronunciation and intonation; and sustained flow of speech in English. The study recommended that cadets maintain an intermittent English learning schedule and have "ground theory lessons" and that flight training discussion of content be taught more extensively in English.

Xiaoning Guo (author of this section) is an English language flight instructor at the Civil Aviation Flight University of China (CAFUC), Sichuan Sheng, China. She teaches English-related courses for CAFUC cadets and has worked on collaborative research projects with colleagues from the United States and members of the International Civil Aviation English Association (ICAEA).

4.3.2. Working with mechanics and aviation maintenance manuals

The works of Drury, Ma, and Marin (2005) and a related report by the same authors (Ma, Drury, & Marin, 2010) have focused on aviation maintenance manuals, particularly on the impact of language errors related to written documentation in aviation maintenance. The researchers targeted non-English primary technicians working in outsourced Maintenance, Repair, and Overhaul (MRO) organizations. Utilizing a survey and a series of open-response questions, just under 1000 technicians in the United States, Latin America, Europe, and Asia were presented with seven different language error scenarios which they then had to resolve. Drury, Ma, and Marin found that language errors occurred frequently in scenarios involving non-English primary workers, but these errors were quickly detected and repaired before they could influence subsequent tasks. The effectiveness of a variety of intervention methods to prevent language error was explored, including increased English literacy skill training, the use of Simplified Technical English (STE) in technical documents, local language glossaries for critical vocabulary, and direct translation of task-cards and manuals. In Asia and Latin America, the potential for language error is highest, as Ma, Drury, and Marin report that technicians from these areas have an average reading level of 5 on the LEXILE scale. Manuals, however, have a LEXILE score of 7.9. Nonetheless, the only intervention methods that the researchers found effective were direct translation of task cards and increased training in English literacy skills. Notably, the use of STE in developing written documentation was not considered an effective strategy to prevent language errors. However, the FAA report did not include any sort of extensive discourse analysis of maintenance documentation that might have revealed specific linguistic factors that pose challenges to non-English primary readers. [See Chapter 5 for a related discussion of this area of research and Chapters 7 and 8 for pedagogy.]

Based on survey data gathered from 172 pilots and 147 line mechanics, a study by Munro, Kanki, and Jordan (2008) examined the influence of various logistical and organizational factors on logbook entries, which are the primary means of communication between pilots and mechanics for the safe operation of mechanical systems on an aircraft. The authors found that corporate pressures to shorten time on the ground between flights lead to an inability for pilots and mechanics to communicate face to face. Furthermore, pilots are not formally trained as to how to compose a logbook entry, and the space allotted for entries is small, leading to terse and often vague, confusing, or uninformative entries. Because these are potential legal and regulatory documents, often scrutinized by officials at the FAA, many mechanics reported worries that what they wrote might be misconstrued or later held against them. More structured guidelines for writing entries, better training for pilots, and redesigned forms that provide greater clarity and specificity are recommended. Holtaway and Jackson (2017) noted that the "leading factor for the FAA initiating Letters of Investigations (LOI) and taking administrative action on Aircraft Maintenance Technicians (AMTs) is failure to follow written procedures" (p. 275). Obviously what isn't understood correctly can't be followed and implemented correctly. To better understand this phenomenon, Holtaway and Jackson conducted a qualitative study investigating AMTs' attitudes and beliefs about technical documentation and their possible reasons for using or not using manuals in the maintenance process. Technicians overwhelmingly noted that low (or challenging) usability was a primary factor influencing their decisions to flout FAA requirements for manual use. The respondents in this study were nearly all English primary speakers who complained more about manuals not representing the way they did maintenance rather than readability, but that begs the question as to whether non-English primary speakers might also indicate that usability and readability influence their use of technical documentation and manuals.

4.4. An annotated bibliography of related aviation English research studies

Table 4.2 provides an annotated bibliography of selected, published language and communication research in global aviation from 2000 to 2016. The table includes full bibliographic citation of the research, a short description of settings and contexts, and keywords or primary research foci explored in the study.

Table 4.2 Language and communication in aviation: An annotated research bibliography

Year	Article	Description	Primary Focus: Keywords
2000	Sexton, J. B. & Helmreich, R. L. (2000). Analyzing cockpit communications: The links between language, performance, error, and workload. *Journal of Human Performance in Extreme Environments*, 5(1), 62–68.	The importance of effective communication on the flight deck is discussed and an application of a new computer-based linguistic method of text analysis is introduced. Preliminary results from a NASA B727 simulator study indicate that specific language variables are moderately to highly correlated with individual performance, individual error rates, and individual communication ratings. Also, language use was found to vary as a function of crew position and level of workload during the flight. Use of the first person plural (*we, our, us*) increases over the life of a flight crew, and CAs speak more in the first person plural than FOs or FEs. Language use in initial flights was associated with performance and error in subsequent flights.	Intra-crew communication; human performance; quantitative and computational analyses.
2002	Corradini, P. & Cacciari, C. (2002). The effect of workload and workshift on air traffic control: A taxonomy of communicative problems. *Cognition, Technology & Work*, 4, 229–239.	Using a corpus of 10 hours of radio exchanges between ATCs and pilots in an Italian airport, this study presents and tests a taxonomy of possible communicative errors during workshift (backward rapid rotation) in varying workloads of ATCs. The taxonomy was able to capture a variety of communicative problems, and the analysis showed that shiftwork and workload significantly affected the ATCs' communicative performance. Communicative errors included non-standard phraseology and expressions and the use of ellipses, redundancies, and the local language, with an increased amount of problems occurring during the nightshift.	ATC communication; aviation psychology; communication and communicative performance; human factors; workload and workshift.

(Continued)

Year	Article	Description	Primary Focus: Keywords
2002	Sullivan, P. & Girginer, H. (2002). The use of discourse analysis to enhance ESP teacher knowledge: An example using aviation English. *English for Specific Purposes, 21*, 397–404.	In this study, a teacher-researcher in an ESP program in a Civil Aviation School in Turkey tape-recorded pilot-ATC communication, made observations in the airport tower, and collected responses to questionnaires and interviews with Turkish pilots and ATCs. Findings indicate that even in a restricted and globally monitored language like Airspeak, local variations exist in greetings, the pronunciation of numbers, and the use of the local language. The authors recommend that findings are used to design and enhance materials for ESP courses.	ESP for aviation; aviation in Turkey; Airspeak.
2003	Brown, N. M. & Moren, C. R. (2003). Background emotional dynamics of crew resource management: Shame emotions and coping responses. *The International Journal of Aviation Psychology, 13*(3), 269–286.	The authors hypothesize that pilots' coping responses to shame (one of Tomkin's nine basic affects) and to the emphasis on professionalism may contribute to crewmembers' occasional failures to communicate concerns to the pilots. Responses from 148 student pilots showed that pilots tend to avoid self-blaming behavior and rely on shame responses of politeness, awkwardness. In addition, the data in this study suggest that pilots are more excitement-hungry than nonpilots, and that they tend to focus instead on excitement, enjoyment, or self-confidence. Based on these findings, this article recommends new training that addresses these avoidance responses to awkward situations in the cockpit.	Affect; emotional dynamics; crew dynamics; pilot personality; training; politeness and sociolinguistic features of discourse.

Year	Article	Description	Primary Focus: Keywords
2003	Helleberg, J. R. & Wickens, C. D. (2003). Effects of data-link modality and display redundancy on pilot performance: An Attentional Perspective. *The International Journal of Aviation Psychology, 13*(3), 189–210.	Using data collected from fifteen certified flight instructor pilots who flew through a Frasca flight simulator, this study examines the effectiveness of auditory, visual, and redundant presentation of ATC information by presenting ATC instructions through synthesized voice, text-link data display, and redundant voice-text formats. After receiving the instructions, pilots' visual scanning was measured as they read back instructions and monitored for traffic. The effectiveness of the three different data link interfaces was evaluated, with results interpreted in terms of attention and working memory in information processing and in terms of pilot's task priority hierarchy. The results revealed that the visual display provided greatest accuracy of communications readback, and was least disruptive of both traffic monitoring and flight path tracking. The auditory-only condition was most disruptive of these tasks, while the redundant display condition provided many of the same benefits as the visual-only display, but was not associated with better performance than that display, and was sometimes inferior to it. Across all conditions, the pilots allocated approximately 60 percent of their visual attention to monitoring the instrument panel and their communications accuracy was degraded by the longer ATC instructions.	Multi-modal information presentation; visual monitoring; attention; working memory.

(Continued)

Year	Article	Description	Primary Focus: Keywords
2004	Rantanen, E. M., McCarley, J. S., & Xu, X. (2004). Time delays in air traffic control communication loop: Effect on controller performance and workload. *The International Journal of Aviation Psychology, 14*(4), 369–394.	The impact of digital technology-induced delays in the pilot-controller communications on ATC's performance and workload was investigated in two experiments manipulating systemic audio delay (AD) and variable pilot delay (PD). Vectoring accuracy (Experiment 1), lateral separation between two aircraft on converging tracks (Experiment 2), and subjective workload (NASA–Task Load Index) were measured. In Experiment 1, the effect of AD was measurable but not statistically significant; PD reduced accuracy and resulted in earlier turn initiations. In Experiment 2, both AD and PD significantly reduced separation between aircrafts. Short communication delays, however, did not substantially degrade controller performance in the experiments.	Pilot-ATC communication; technology-induced delays; audio delay; pilot delay.
2006	Morris, C. H. & Leung, Y. K. (2006). Pilot mental workload: How well do pilots really perform?, *Ergonomics, 49*(15), 1581–1596.	This study investigated the effects of increasing mental demands on various aspects of aircrew performance. In particular, the robustness of the prioritization and allocation hierarchy of aviate–navigate–communicate, commonly used within the aviation industry, was examined. A total of forty-two trainee pilots were divided into three workload groups (low, medium, high) to complete a desktop, computer-based exercise that simulated combinations of generic flight deck activities: flight control manipulation, rule-based actions and higher level-cognitive processing, in addition to ATC instructions that varied in length from one chunk of auditory information to seven chunks. It was found that as mental workload and auditory input increased, participants experienced considerable difficulty in carrying out the primary manipulation task and a decline in prioritization. Pilots also had difficulty comprehending more than two auditory stimuli when under a high mental workload.	"Aviate-Navigate-Communicate;" auditory input; instruction comprehension.

Year	Article	Description	Primary Focus: Keywords
2007	Sharples, S., Stedmon, A., Cox, G., Shuttleworth, A. N. T., & Wilson, J. (2007). Flight deck and air traffic control collaboration evaluation (FACE): Evaluating aviation communication in the laboratory and field. *Applied Ergonomics, 38*, 399–407.	This paper presents a research program that utilized both laboratory- and field-based assessments to examine the way in which datalink and free flight may affect the communication and collaboration between pilots, ATCs, and other actors and artifacts in the flight deck–ATC joint cognitive system. The methodological approach is discussed and an overview of the results from these studies is presented, as well guidance is provided as to the likely situations in which this new technology is most likely to be successfully applied.	Pilot-ATC communication; datalink; freeflight; technology and communication.
2008	Howard, J. W. (2008). "'Tower, am I cleared to land?'": Problematic communication in aviation discourse. *Human Communication Research 34*, 370–391.	This study examined more than 15 hours of pilot-ATC dialogue across 15 US airports to analyze problematic communication in pilot-ATC interaction. The transcribed data included a total of 34 ATCs, 270 pilots, and 1,799 turns of talk. Analyses revealed that (a) communication problematics manifested in pilot turns more than ATC turns, (b) higher amounts of information led to increased problematic communication in the subsequent turn, and (c) linguistic violations of ATC protocol increased problematic communication in the subsequent turn. Partial support was found for the effect of frequency congestion on problematic communication, but not for airport size. The discussion addresses protocol deviations and system constraints for problematic communication. Applications for the findings in pilot-ATC radio interaction are also suggested.	Pilot-ATC communication; problematic communication.

(Continued)

Year	Article	Description	Primary Focus: Keywords
2008	Prinzo, O. V. & Hendrix, A. M. (2008). Pilot English language proficiency and the prevalence of communication problems at five U.S. air route traffic control centers. US Department of Transportation (Final Report).	Although ICAO requires its contracting states to ensure ATCs are proficient communicators in English, data are lacking in communication problems attributable to errors in readback-hearback loops. This report examines communication breakdowns in 50 hours of air-ground transmissions provided by five ARTCCs from 832 aircrafts and 4,816 pilot transmissions, scoring each transmission for accuracy and encoding English proficiency using ICAO Language Proficiency Scale. Call signs were classified by aircraft registry, resulting in three groups: Foreign-English, Foreign-Other, and US-English. Findings demonstrated that 23 percent of the transmissions contained one or more communication problems, 75 percent among Foreign-Other aircrafts, and 29 percent involving US-English aircrafts. Communication problems included pronunciation and fluency with Foreign-Other aircrafts and fluency in US-English aircrafts. The authors recommend language proficiency beyond the minimum requirement to avoid communication problems.	Air traffic control training and administration; ATC communication; communication situations; communication problems.
2009	DeMik, R. J. (2009). Text communications in single-pilot general aviation operations: Evaluating pilot errors and response times. *International Journal of Applied Aviation Studies, 9*(1), 29–42.	To test the text-based Controller-Pilot Data Link Communications (CPDLC) technology's effect on human performance in the single-pilot general aviation work environment, Instrument-rated pilots (N = 30) were tested on both a series of CPDLC text command tasks and conventional voice command tasks in a simulated flight environment. Results revealed a statistically significant decrease in both errors in pilot recall/execution and response times in moving from the conventional voice ATC commands to the CPDLC text commands. Results lend support for a text-based communications technology that extends pilot's working memory, thereby improving both accuracy and efficiency in performing flight tasks to complex commands.	Controller-Pilot Data Link Communications (CPDLC); text commands; communications technology.

Year	Article	Description	Primary Focus: Keywords
2010	Alderson, J. C. (2010). A survey of aviation English tests. *Language Testing* 27(1), 51–72.	Based on an internet search for ATC language tests, the Lancaster Language Testing Research Group (LLTRG), a group commissioned by the European Organization for the Safety of Air Navigation (Eurocontrol) conducted a validation study of the development of a test called ELPAC (English Language Proficiency for Aeronautical Communication). The study included a questionnaire based on the Guidelines for Good Practice of the European Association for Language Testing and Assessment (EALTA), and LLTRG elicited twenty-two responses from various organizations. Little evidence was found to attest to the quality of these tests: there was high variability in the tests, which they hypothesize represents variation in test quality and low awareness of appropriate procedures for test development, maintenance, and validation. Therefore, LLTRG concludes that there is little consistency in the meaningfulness, reliability, and validity of current aviation language tests, and suggests that language tests be more closely monitored.	Aviation language tests; Eurocontrol; test variability; language test validation.
2010	Brown, L. J. & Rantz, W. G. (2010). The efficacy of flight attendant/pilot communication in a post-9/11 environment: Viewed from both sides of the fortress door. *International Journal of Applied Aviation Studies*, 10(1), 227–248.	Intra-crew communication has been complicated due to physical and procedural changes after September 11, 2001, and ineffective communication and coordination has been cited as the cause of many incidents. To address gaps in research on pilot-flight attendant communication, this study collected responses to a self-reporting survey from 112 pilots and 230 flight attendants worldwide. Results indicate the dire need for improved flight attendant/pilot communication, coordination, and several recommendations for improvements are discussed.	Pilot-flight attendant communication; intra-crew communication; post-9/11 environment.

(Continued)

Year	Article	Description	Primary Focus: Keywords
2010	Prinzo, O. V., Campbell, A., Hendrix, A. M., & Hendrix, R. (2010). U.S. airline transport pilot international flight language experiences Report 2: Word meaning and pronunciation. Defense Technical Information Center, Ft. Belvoir, VA 22060; and the National Technical Information Service, Springfield, VA 22161.	**Report 2:** After the Air Navigation Commission by ICAO directed to strengthen language requirements in 1998, Member States ensured that ATC personnel and flight crews would be proficient in conducting and comprehending radiotelegraphy communications in English. Furthermore, ICAO encouraged its members to document their English Language Proficiency (ELP) requirements and implementation plans in 2008. To better understand the language difficulties in international operations, a group of forty-eight pilots evenly recruited from American, Continental, Delta, and United Airlines met with interviewers to discuss their experiences flying into various countries. This report presents a compiled narrative of their responses and comments to questions 24–30. There were eight major issues addressed by pilots, including: the ATC's accented English in routine operations and its effects on flight time currency; the effect on comprehension caused by the lack of standardized phraseology and pronunciation, as well as variations in pronunciation; the poor quality radio and transmission; the need for training for pilots going to a foreign country; and technological advancements. Ten recommendations are provided to address pilots' concerns.	Air Navigation Commission (ANC); English language proficiency; narrative; focus groups; accented English; phraseology; pronunciation training.

Year	Article	Description	Primary Focus: Keywords
2010	Prinzo, O. V., Campbell, A., Hendrix, A. M., & Hendrix, R. (2010). U.S. airline transport pilot international flight language experiences, Report 3: Language experiences in non-native English speaking airspace/airports. Defense Technical Information Center, Ft. Belvoir, VA 22060; and the National Technical Information Service, Springfield, VA 22161.	**Report 3:** Using the same data from Report 2 [above], this report presents a compiled narrative of pilots' responses and comments to questions 31–38. There were nine major points made by pilots, including: the positive impact on pilots of traveling to non-native English-speaking countries; the variance of English proficiency and voice characteristics in individuals and countries, as well as in different-from-normal procedures; the negative effects of hearing multiple languages on the radio, including reduced situational awareness, loss of radio protocol, and diminished flight deck operations; and the positive effect of technology to avoid complications.	Pilot-ATC communication; non-native English-speaking ATC; English language proficiency; radiotelephony; technology and aviation.
2010	Prinzo, O. V., Campbell, A., Hendrix, A. M., & Hendrix, R. (2010). U.S. airline transport pilot international flight language experiences, Report 4: Non-native English-speaking controllers communicating with native English-speaking pilots. Defense Technical Information Center, Ft. Belvoir, VA 22060; and the National Technical Information Service, Springfield, VA 22161.	**Report 4:** This report presents a compiled narrative of pilots' responses and comments to questions 39–45. There were six major points brought up by pilots, including: their concern for non-native English-speaking controllers in high workload condition; their varying strategies for different ATC language proficiency; their descriptions of their communications with ATCs from various language backgrounds; the negative effects of switching languages and language barriers; their compensation for reduced situational awareness. Sixteen recommendations are made to improve communication practices, including suggestions such as developing standard language tests, providing training on emergency and non-routine situations, and simulations of conversational English.	Pilot-ATC communication; code-switching; negative effects.

(Continued)

Year	Article	Description	Primary Focus: Keywords
2010	Prinzo, O. V., Campbell, A., Hendrix, A. M., & Hendrix, R. (2010). U.S. airline transport pilot international flight language experiences, **Report 5:** Language experiences in native English-speaking airspace/airports. Defense Technical Information Center, Ft. Belvoir, VA Service, Springfield, VA 22161.	**Report 5:** [follow-up reports from Report 4, above]. There were eight major points brought up by pilots, to which the authors present five recommendations: the adoption of a standard dialect and phraseology in ATC communications, and the discouragement of local language; additional training and instruction in voice production and articulation as they relate to ATC communication; the definition of an optimal speaking rate and research to determine the optimal rate for speakers from different backgrounds.	Pilot-ATC communication; dialect; phraseology; training and instruction; future research.
2011	Prinzo, O. V., Campbell, A., Hendrix, A. M., & Hendrix, R. (2011). U.S. airline transport pilot international flight language experiences, **Report 6:** Native English-speaking controllers communicating with non-native English-speaking pilots. Defense Technical Information Center, Ft. Belvoir, VA 22060; and the National Technical Information Service, Springfield, VA 22161.	**Report 6:** This report presents a compiled narrative of pilots' responses and comments to questions 54–59. Five recommendations are presented: there should be additional research to determine the optimal speech rate for ATC messages, how ATCs communicate non-routine situations, and the effects of the absence of party-line information on situational awareness and safety; ATC messages must be delivered using standard ICAO terms and phraseology; and pilots should receive taxi and route clearances and modifications as standalone messages.	Pilot-ATC communication; future research; phraseology.

Year	Article	Description	Primary Focus: Keywords
2012	Ansdell, M. (2012). Language protocols in international human spaceflight: Time for a common tongue? *Space Policy, 28*, 2–6.	The author suggests that language protocols for space exploration be examined, as the language of prior spaceflight missions was dictated by political realities and resulted in difficulties. Furthermore, the author suggests that the international space community should adopt a single-language protocol but consider major languages other than English.	International language protocol; space language protocol.
2012	Comitz, P. (2012). The Boeing Company, advanced air traffic management: A domain specific language for aviation domain interoperability. Integrated Communications Navigation and Surveillance (ICNS) Conference April 24–26, 2012.	The purpose of this work is to promote and advance interoperability between aviation information systems. Complex and error-prone communication is often caused by the limited interoperability and collaboration between different organizations, many of which use different aviation systems with differing representations and distribution policies for the same data and messages. The complexity and variation in information-processing environments create a barrier between domain practitioners and engineers, contributing to failures such as the FBI's Virtual Case File.	Interoperability; communication; aviation information systems.

(Continued)

Year	Article	Description	Primary Focus: Keywords
2012	Cutting, J. (2012). English for airport ground staff. *English for Specific Purposes, 31*(2), 3–13.	To design a multimedia course for English language learners seeking employment in European airports, the structural-functional analysis of four trades (security guards, ground handlers, catering staff and bus drivers) found position-specific language. Across all four, the present simple and clausal ellipsis abounded. However, will-future was mostly used by security guards, ground handlers, and bus drivers when referring to regulations and offering their own action. Direct orders were given mainly by security guards and bus drivers, to enforce the law and keep passengers moving. Passenger complaints were met by security guard pseudo-apologies, but bus drivers rarely apologized. Catering staff and bus driver dialogues featured the polite would, can and could in offers and requests, accompanied by greetings, farewells, pleases, thank yous and formal address forms. Security guards used hedges to mitigate their threat to passengers while ground handler routines allowed little time for interactional softeners.	Airport ground staff; dialogue analysis; forms and functions of workplace discourse.
2013	Dalto, J. D., Weir, C., & Thomas, F. (2013). Analyzing communication errors in an air medical transport service. *Air Medical, 32*(3), 129–137.	Using 278 quality assurance reports, the authors classified and mapped communication errors using Clark's communication level hierarchy. They found sixty-four errors in fifty-eight reports, with levels 1 (66 percent) and 4 (33 percent) being the most common. Less than one third of these errors were identified by the reporting staff. An air medical communication ontology is necessary to improve the recognition and analysis of communication errors.	Air medical communication errors; assurance reports; Clark's communication level hierarchy.

Year	Article	Description	Primary Focus: Keywords
2013	Ford, J., Henderson, R. & O'Hare, D. (2013). Barriers to intra-aircraft communication and safety: The perspective of the flight attendants. *The International Journal of Aviation Psychology, 23*(4), 368–387.	To identify perceived barriers to teamwork between pilots and flight attendants, this study used the nominal group technique (NGT) to gather data from 100 flight attendant through focus groups. The NGT methodology generated barriers and solutions for effective communication, including: the locked flight deck door and interphone protocols; "sterile cockpit" standard operating procedures (SOPs); preflight briefings; knowledge of basic aircraft terminology; debriefings after incidents; and contractual differences in hotels, meals, and allowances. The authors recommend joint crew resource management (CRM) training and that the solutions presented here be written into CRM and emergency procedures training, as well as training for pilots and flight attendants to better understand the other's role.	Pilot-flight attendant communication; communication barriers; teamwork; nominal group technique.
2013	Kim, H. (2013). Exploring the construct of radiotelephony communication: A critique of ICAO English testing policy from the perspective of Korean aviation experts. *Papers in Language Testing and Assessment, 2*(2), 103–110.	This study examined perspectives of and resistance by Korean aviation personnel toward ICAO policy. This included examining how Korean pilots perceived the representation of the requirements of radiotelegraphy communication in tests and ICAO policy, and the important qualities of radiotelephony discourse. Most of the feedback received focused on the professionalism and adherence to prescribed conventions. Participants perceived the main tendency of NES pilot discourse to be the use of general English, including in an unnecessarily wordy manner.	Radiotelephony; professionalism; conventions; resistance; Korean aviation setting.

(Continued)

Year	Article	Description	Primary Focus: Keywords
2014	Katerinakis, T. (2014). Aviate, navigate, communicate: Silence, voice and situation awareness in aviation safety. Unpublished PhD Dissertation, Drexel University.	This dissertation examines the mediated environment of the flight deck by infusing communicative components to both human factors and systems theoretic approaches to aviation. The author views the discursive space of a cockpit as one consisting of actors in specific social roles and under time-critical situations. Interlocutors must keep track of demands, the system in its entirety and their position, and the consequentiality of their decisions.	Mediated environment; human factors; silence and voice awareness.
2014	Mayin, T. J. & Roth, W. (2014). A holistic view of cockpit performance: An analysis of the assessment discourse of flight examiners. *The International Journal of Aviation Psychology*, 24(3), 210–227.	The authors used a discursive psychology approach to analyze the discourse of flight examiners in seven extended interviews about performance aspects. In their discourse, flight examiners presented cockpit performance holistically, even though it manifests in different ways. Six main discourse repertoires are identified in examiners' discourse about flight deck performance, each of which has between three and five identifiable sub-dimensions. Case studies show the connectedness and inter-determination of the six main repertoires for talking about what pilots do and how they do it.	Discursive psychology; flight examiners; flight deck repertoires.
2016	Tanguy, L., Tulechki, N., Urieli, A., Hermann, E., & Raynal, C. (2016). Natural language processing for aviation safety reports: From classification to interactive analysis. *Computers in Industry*, 78, 80–95.	This paper examines aviation incident reports and presents tools that aim to provide better access and analyze data from them in order to inform safety experts. The authors focus on using NLP techniques to analyze the reports that, despite being written in plain language, show a wide variety of linguistic variation. This variation includes acronyms and prose in multiple languages across a large quantity (approximately 600 per month for a large airline company), which requires advanced NLP and text mining to extract useful information.	Aviation language; document classification; NLP; safety reports; text mining.

4.6. Future research

Aviation English research is increasing and expanding in scope, both in terms of the areas of focus and the collaborative initiatives discovered and utilized to increase the body of knowledge related to aviation communication and factors influencing it. There are many promising signs of research support as governmental organizations and airlines comprehend the need and show interest in assisting language-based research and in developing innovative training programs for pilots, ATCs, maintenance workers, flight crew, and ground staff.

Estival (2018) highlighted what she sees as immediate future research directions, focusing especially on data analyses of live recordings, for example, data from LiveATC recordings of interactions from international airports such as those in Sydney, Hong Kong, Tokyo, and Los Angeles; the analysis of flight simulator data to possibly discover correlations with pilot communication errors that may be identified from audio recordings; and the development of automated training tools, using automated speech recognition (ASR) and text-to-speech (TTS) software that might later be deployed operationally. For Estival, the clear contributions of these initiatives and other related research include heightened awareness on the part of both ATCs and pilots, both NES and NNES, of potential communication problems that are particularly problematic but not, as yet, adequately addressed in training and proficiency assessments, and to provide support for practicing RTF in realistic aviation contexts outside the cockpit. The ultimate goal of all of these efforts will certainly increase the competence and confidence of aviation personnel and consequently air travelers' safety, as well.

Chapter Co-Authors

Justin C. Taylor received his master's degree in Applied Linguistics and ESL from Georgia State University. He has experience in teaching English to international students in the United States, Turkey, and China, and he has taught, supervised, and advised American undergraduate and international exchange students at GSU. His primary research interests are sociolinguistics, intercultural communication, and ESL/EFL pedagogy.

Rachelle Udell is a doctoral student in Applied Linguistics and ESL at Georgia State University. Her research specializations include ESP—Aviation English, L2 literacy, corpus-based discourse analysis, and intercultural communication. She also focuses on various language policy and planning implications of Aviation English, particularly on written documents such as technical manuals and texts from Controller Pilot Data Link Communications (CPDLC), a two-way, text-based communication between pilots and air traffic controllers, freeing up one-way radio channels for more urgent communication.

5

The Language of Aviation: Corpus-Based Analysis of Aviation Discourse

With Genggeng Zhang

5.1. Introduction

Corpus linguistics (CL) is a research approach to the empirical investigation of language variation and use, primarily utilizing a quantifiable collection of "texts" known as a corpus (*corpora*, plural) and various computer-based tools. These tools are developed to process and analyze corpora automatically and electronically, producing data and results that have, arguably, much greater generalizability and validity and subsequent utility than would be feasible using traditional linguistic and discourse analytic approaches (Biber, Reppen, & Friginal, 2010). When applied to analyzing spoken and written discourse in the aviation industry, CL systematically provides relevant and meaningful data—frequency distributions and actual patterns of English vocabulary and grammar in natural, authentic contexts (e.g., pilot and air traffic controller communication or maintenance manuals from airline manufacturers). Corpora and CL approaches support the view that language, in general, is systematic and can be described and examined using empirical, frequency, and pattern-based approaches (Friginal, 2018). Specifically, for the aviation industry, CL facilitates practical investigations of aviation English domains that can be extensively analyzed and interpreted, and then taught to second or foreign speakers of the language (i.e., ESL/EFL speakers).

Collected texts form a *corpus*, which refers to a compilation of written and spoken language stored on a computer. Here, the term **text** includes transcription of speech from actual audio or video recordings of interactions. Some of these spoken texts may have been annotated or marked to include oral and auditory characteristics of spoken discourse (e.g., tone, pitch, loudness), but there are certainly other speech features that are not as easily captured in electronic corpora. Pilot non-native accent and intonation and related speech acts and pragmatic markers of sarcasm or humor, for example, cannot be directly ascertained from corpora.

To emphasize, a linguistic corpus is, by definition, computerized and searchable by computer programs (Friginal & Hardy, 2014a). A corpus of aviation English texts can be created and compiled, as long as there is a logical and linguistically principled design that guides the collection process. In considering what may constitute a sound aviation English corpus, the following definitions of corpus/corpora from corpus linguists over the past few years may be helpful:

> ... computer corpora are rarely haphazard collections of textual material: They are generally assembled with particular purposes in mind, and are often assembled to be (informally speaking) *representative* of some language or text type. (Leech, 1992)
>
> A corpus is a collection of pieces of language text in electronic form, selected according to external criteria to represent, as far as possible, a language or language variety as a source of data for linguistic research. (Sinclair, 2005)
>
> Corpora may encode language produced in any mode—for example, there are corpora of spoken language and there are corpora of written language. In addition, some video corpora record paralinguistic features such as gesture (Knight et al., 2009) and corpora of sign language have been constructed (Johnston & Schembri, 2006; Crashborn, 2008). (McEnery & Hardie, 2012, p. 3)
>
> ... a corpus can be briefly defined as a systematically designed electronic collection of naturally occurring texts [...] Researchers compile corpora and search for existing constructs of written or speech patterns identified as relevant and measurable. A corpus provides the opportunity to measure tendencies and distributions across registers (and genres) of language. (Friginal et al., 2017)

From the definitions above, an aviation English corpus—systematic compilation of naturally occurring language in the aviation industry—serves as a primary dataset for researchers interested in describing and analyzing linguistic forms, functions, and variation in this domain. For instructors and trainers, it is an important source of linguistic information that can inform and direct language training and assessment programs. The clear argument for this approach is that actual usage gleaned from corpora is important in analyzing, teaching, and evaluating aviation English.

A key element of the definitions above, especially Leech's (1992) and Sinclair's (2005) definitions of a corpus, is that texts *represent* a language or a language variety (also referred to as *register*). Biber (1993) defines **representativeness** as "the extent to which a sample includes the full range of variability in a population" (p. 243). In a more general sense beyond CL, representativeness refers to the idea that one can collect a smaller sample than the population as a whole, but that that smaller sample could show as much variation and range in the subset as in the overall population. This description is very relevant for aviation English researchers as it is still quite difficult to collect a fully realized, large-scale corpus of language across the board in the industry.

The representativeness of a corpus can be considered both contextually and linguistically. Contextually, a corpus of aviation English should include the full range of existing registers. As different situations in which English is used in the industry affect the way it is actually utilized across contexts, those different sub-registers, including groups of speakers or participants, need to be included in order to fully understand the variety or domain as a whole. Linguistically, an aviation English corpus could be representative if it includes the full range of different lexical and grammatical features also characterizing these contexts. In the following sub-sections of this chapter, and also in Chapter 6, exploratory corpora of airplane maintenance manuals and pilot-ATC interactions are analyzed using a range of CL tools and statistical tests. These corpora followed a principled design and are collected to address the concept of *balance* and representativeness, but clearly they are still relatively limited in size and scope. Results presented here, however, are very promising, with interesting findings and applications to pedagogy and in helping to provide a model for corpus design of a fully representative corpus of language in aviation.

5.2. CL, sociolinguistics, and the language of aviation

The aviation English register: Biber, Conrad, and Reppen (1998) make use of the term **register** to describe a situationally defined category of speech and writing. A register distinction of spoken texts, for example, can cover sub-registers such as face-to-face interaction, telephone interaction, and video calls (e.g., Skype calls or mobile "face-time" calls). Corpora representing these three sub-registers could be collected and transcribed. These sub-registers are differentiated by the medium and contexts, which can certainly influence the use of a whole range of linguistic features. Register variation, therefore, is primarily based on these contextual differences. A researcher may be able to control register comparisons and the situations that define them. The term "register" has often been used interchangeably with *genre*. A genre perspective in research, however, is more interested in the conventional structures that are used to create an entire text or discourse, or a section of a text, such as research article introductions or the abstract from a research article (Flowerdew, 2005; Friginal & Hardy, 2014a).

In researching the linguistic characteristics of aviation English as a sub-register of spoken and written discourse, a model situated in the intersection of CL and the field of **sociolinguistics** (i.e., corpus-based sociolinguistics) could serve as a theoretical guide. Friginal and Hardy (2014a) define sociolinguistics as:

> The study of variation in language *form* and *use* that is associated with social, situational, attitudinal, temporal, and geographic influences. (p. 4)

Studies in sociolinguistics have investigated why and how individuals across varying backgrounds speak and write differently. They have also explored the effects of various aspects of society, which include societal expectations, cultural norms and traditions, and historical backgrounds on the way language is used. By examining the influence of societal variables on language use and its resulting structure, aviation English researchers can better comprehend and describe the reality that everyday language even in this relatively "controlled" setting is remarkably varied and influenced by numerous factors. The implication of this is that no one pilot or ATC speaks the same way all the time, and speakers and writers in this particular domain of communication constantly exploit the nuances of the languages they speak and write for a wide variety of purposes in a variety of contexts.

This recognition of variation in language use suggests that everyone must see language as not just some kind of abstract object of study (Meyerhoff, 2004) but a dynamic and evolving realm. Arguably, English for global aviation does not need to follow one particular model of language, especially the native-speaker model. In a broader sense, this concept emphasizes that the discourse of aviation is simply different from other types of professional, workplace discourse (in English). Even more specifically, it also reflects that the language of pilots representing different first language backgrounds, airlines, length of service in the industry, or genders will show noticeable variation. This is also the case with ATCs serving in various airports all over the world. Language, as it is, is pragmatic, practical, evolving, and unique to individuals or groups of connected

Table 5.1 Linguistic and contextual variables of aviation English explored using corpora (adapted from Friginal & Hardy, 2014a)

	Variables	Sample Research Ideas: Spoken Aviation Discourse	Sample Research Ideas: Written Aviation Discourse
	Linguistic		
1	**Sounds, words, and grammatical features of a language**—includes a range of differences in the pronunciation of sounds, intonation of utterances, and the use of words and phrases (and also dysfluent markers of speech), and grammatical structures of language.	Compare the frequencies of a range of vocabulary features (standard phraseology and common language features) across groups of international pilots and determine if these frequencies correlate with or affect interlocutor comprehension.	Describe and quantify the readability of airplane maintenance manuals relative to the reading ability (in English) of technicians (see sample study below).
2	**Discoursal features**—includes spoken and written characteristics of style, formality or informality of discourse, and textual structures (e.g., use of cohesive devices in writing; interruption, latching, or overlaps in face-to-face conversation).	Further explore spoken norms or mannerisms of international pilots as they communicate in various take-off, in-flight, and landing situations; collect corpora that will be coded or annotated with discoursal features of speech such as interruption or overlaps.	Extract and analyze written electronic commands and messages from emergency announcement systems; analyze the distributions of connectors or linking adverbials in maintenance manuals compared to other technical manuals and written academic texts.
3	**Pragmatic features**—includes spoken and written expressions of politeness in language, stance and hedging, the use of respect markers or cuss words, and features of agreements and disagreements in interactions.	Compare how native English and non-native English speakers (pilots/ATCs) make use of grammatical politeness markers (e.g., *thanks, appreciate, sorry, apologize*) across conversational contexts; establish the relationship that may exist between the distribution of these markers and speaker demographics, role-relations, and transactional tasks.	Study directness and features of accuracy in writing for technical manuals and other written sub-registers of the aviation industry; explore if there are differing reactions to how these manuals are understood or interpreted across cultural groups. Explore noticeable and measurable representations of standard phraseology in written aviation English texts.

(Continued)

	Variables	Sample Research Ideas: Spoken Aviation Discourse	Sample Research Ideas: Written Aviation Discourse
4	**Specific communicative features**—includes spoken and written manifestations of friendliness, affection, loyalty, or disgust; various speech acts (e.g., requests, commands, and declarations); pauses, backchannels, greetings and leave-takings; and visual representations of attitude, political positions, and personal/group opinions and biases in print media.	Establish quantifiable features of ATC commands provided to pilots of differing cultural and language backgrounds; measure length of pause and frequency of filled-pauses by international pilots and determine if these relate to interlocutor response or attitude.	Extract various lexical bundles or n-grams of written sub-registers of aviation technical and accident reports; interpret their functions and intended primary goals (see below from Section 5.4.2).
5	**Paralanguage features**—includes pitch and volume in speech and non-verbal elements of language such as silence, gasp, laughter in conversations; paralanguage may also include the use of visuals (e.g., pictures, colors, signs and signage), emoticons, or punctuation marks in writing.	Find ways to code or annotate corpora for intonation, volume, or manifestations of second language accent in speech, e.g., The Hong Kong Corpus of Spoken English—Prosodic (Cheng, Greaves, & Warren, 2008); explore how these features vary across spoken or interactional contexts.	Code and annotate visuals (colors and design elements) typically employed in technical manuals compared to other types of professional/technical texts.
	Contextual		
6	**Social**—speaker/writer demographic information such as gender and sexuality, age, occupation, educational background, annual income, group networks (traditionally, social networks—not referring to the internet or social media applications such as Facebook, Twitter, or Instagram), social class or social status.	Notable speaker categories that could be prioritized for comparison: **roles** (pilots vs. ATCs); **language background** (NNES vs. NS); **age** and **gender; years of flying experience (pilots)**.	**Text categories**: technical/maintenance manuals vs. other written documents.

Variables	Sample Research Ideas: Spoken Aviation Discourse	Sample Research Ideas: Written Aviation Discourse
7	**Situational**—various communication contexts and **registers**; speech events such as conversation, interview, or broadcast.	Explore: **settings or locations** (e.g., international pilots landing in the United States; US-based pilots or carriers landing in China; emergency vs. non-emergency communication settings).	Compare: Various sections of a particular text or aviation document (e.g., maintenance manuals, memos, reports) and how specific linguistic and design features may vary.
8	**Attitudinal and Relational**—speaker/writer perceptions and attitudes (including prejudice), identity and identity construction, power, relationships and roles, solidarity.	Extract, document, and interpret pilot attitudinal reactions to ATC commands and directions; explore how international pilots use politeness and respect markers in communicating with, especially, US-based ATCs.	(Potentially limited options here for written technical documents) Explore other written texts of aviation discourse, including those related to advertising, job announcements, and social media.
9	**Temporal**—time periods (e.g., "real time" and "apparent time" studies), change in societal and cultural perspectives over time, major historical events including influences from wars, natural calamities, and migration patterns over time.	Explore interactional databases across time periods (e.g., accident reports); collect a temporal and/or a monitor corpus of various spoken interactions in the aviation industry.	Compare written structures and grammatical features of English technical manuals across time periods and manufacturing companies.
10	**Geographic**—particular locations, geographic regions, and boundaries.	Explore various location-based spoken interactions on the distributions of lexico-syntactic features of English.	Compare, for example, Airbus and Boeing texts from a geo-political perspective, relative to the role and features of English.

(Continued)

Variables	Sample Research Ideas: Spoken Aviation Discourse	Sample Research Ideas: Written Aviation Discourse
11 **Other Societal Variables**—more specific personality and cognitive factors, sociological distinctions; "uncommon" or new/emerging societal variables, particularly influenced by the internet; human-nonhuman and machine-mediated communication; technology-based variables (e.g., use of telecommunication devices, gaming devices and gaming culture).	Code and annotate corpora based on human factors and individual demographic information of speakers (e.g., pilots and ATCs, flight attendants; roles and relationships; recorded events and situations, and others).	Further collect written sub-registers of aviation discourse, especially those that could be categorized or coded for themes and foci; explore emerging texts such as the use of short message services (SMS) and other technology-based devices.

individuals. Concluding remarks and generalizations can be formulated relative to these variations and their practical implications, and, additionally, answers to questions such as: How do these variations influence policies or attitudes? How could these patterns be taught effectively to better ensure complete and accurate comprehension of communication? How do pilots and ATCs address linguistic differences to make sure that their reactions are accurate or constructive? How have linguistic patterns and conventions changed over time in the industry? These and many other related questions could be answered by utilizing multiple, traditional research approaches in sociolinguistics and, perhaps, may be best described and interpreted following a CL research paradigm.

Table 5.1 provides an overview of the important variables of aviation English that could be explored using a corpus-based sociolinguistic model proposed by Friginal and Hardy (2014a) and Friginal (2018). The main register of aviation English is first divided into two primary sub-registers: spoken and written, and each of these sub-registers is further described across various linguistic and contextual features. The goal here is to identify research ideas and foci that could be developed into concrete research questions answerable by corpus-based approaches. The table also serves to organize and plan various corpus collection projects and subsequent corpus-based analyses.

Exploring the aviation English register from a corpus-based sociolinguistic model, in general, produces data that may indeed show systematic and habitual forms, patterns, and functions of speech and writing in the industry. Although these data from corpora offer measurable descriptions of texts and speaker/writer groups, researchers and subsequent consumers of this information must still functionally **interpret** these corpus-based findings as accurately and consistently as possible. Extensive knowledge of ICAO literature, related approaches, and awareness of the clear limitations of corpus and computational tools must always be considered in the analysis of this data. Interpretive techniques refined by ethnographers and discourse analysts over the years are certainly invaluable. Thus, corpus approaches could certainly be used with qualitative and discourse analytic methods typically employed by industry trainers and analysts, and corpora and frequency data can be statistically tested to determine whether or not a consistent, significant pattern exists.

5.2.1. Corpus tools and related analytical approaches

Finally, in this section, it is worth noting that both the number and utility of available corpus tools that can facilitate the extraction of data identified in Table 5.1 have grown since their initial development. There is now relatively easy access to freely available concordancers, part-of-speech (POS) taggers and parsers, and data extractors that automatically process huge volumes of texts. Corpus findings have gradually explained or defined the functional parameters of linguistic variation based on frequencies and statistical patterns of usage across writer/speaker demographics—together with tools that help in visualizing existing relationships and distributions. Visualizing language-based data easily begins with corpus-based frequencies, which can be transformed into figures or images. From simple bar graphs or histograms to more complex, online interactive semantic maps, corpus tools have produced excellent visual representations of language and communication in context (Friginal, 2018). At the same time, an increasing number of sociolinguistic and language in the workplace corpora have been collected and many linguists, including those who work in aviation, have now progressively utilized corpus-based data and approaches in their own research.

Extracting frequencies of linguistic features from corpora is a basic type of analysis in corpus-based research. Questions such as "What are the most frequently used words by international pilots in pilot-ATC interaction?" or "What are the most common lexical verbs used by NES and NNES ATCs during take-off and landing interactions?" are relatively easy to obtain from relevant corpora. The former simply requires running the word list function of a concordancer such as the freeware *AntConc* (Anthony, 2018) or *WordSmith Tools* (Scott, 2014, available for purchase) while the latter will first require a corpus that is tagged or annotated for POS; that is, the aviation research will have to utilize a POS-tagger to obtain the frequency of the most common lexical verbs—these are meaning-carrying, one-word verbs, such as *sing, talk, think*, or *find* and their lemmas—in the corpus (Friginal, 2018). Concordances can also be utilized to identify the different usages of a key or content word, examine word relationships, explore the distribution of key terms and phrases, and create a list of multiword units. All of these additional features can be produced immediately from *AntConc* [download *AntConc* from: <http://www.laurenceanthony.net/software/antconc/> and the resulting concordance lines can be saved for additional qualitative coding and analyses. The corpus approach also provides the determination of statistically significant word combinations, such as collocations and word chunks. Prediction models of what might follow or precede a word, a noun, or a verb can be measured based on their expected frequencies from corpora.

5.3. Corpus-based analysis of aviation industry discourse

Specialized corpora for specific contexts such as English as a *Lingua Franca* (e.g., the Vienna-Oxford International Corpus of English or VOICE), the Augmentative and Alternative Communication (AAC) User and Non-AAC

User Workplace Corpus (ANAWC) (Pickering & Bruce, 2009), and the Nurse-Standardized Patient Corpus (NSP Corpus) (Staples, 2015) represent actual spoken interactions across workplaces. These types of corpora have already been thoroughly explored and analyzed, thereby providing available and useful models for the similar linguistic examination of spoken interactions in the aviation industry. Several workplace corpora have focused on business and office-based interactions that produce speech events also common in the aviation industry. Hence, there are many opportunities to compare data from these corpora with those from aviation. The Cambridge and Nottingham Business English Corpus (CANBEC) is sub-corpus of the Cambridge English Corpus (CEC), covering several business settings from large companies to small firms, formal meetings and presentations, and lunchtime or break room conversations. Similar to this is the American and British Office Talk (ABOT) corpus, mainly focusing on "informal, unplanned workplace interactions between co-workers in office settings" (Koester, 2010, p.13). The ABOT corpus has been used to study the performance of communicative functions in the workplace using speech acts and relational sequences (Pickering, Friginal, & Staples, 2016).

Workplace corpora in English collected outside the UK and the United States include, as previously mentioned in this chapter, The Hong Kong Corpus of Spoken English (prosodic) (HKCSE) and the Language in the Workplace Project (LWP) from New Zealand. The LWP has transcriptions of office interactions in government organizations, small business, hospitals, IT organizations, and publishing companies. An increasing number of research studies have been published using the LWP since the 1990s, focusing on topics such as analyses of cross-cultural pragmatics, gender and ethnicity, humor and small talk, and speech acts such as directives from multiple discourse perspectives (e.g., Holmes, 2009; Marra, 2012; Stubbe et al., 2003; Vine, 2017). The HKCSE (prosodic) (Cheng, Greaves & Warren, 2008) includes various types of formal and informal office talk, presentations, conference calls, and service encounters in the hotel industry. It qualifies as an "intercultural corpus," with interlocutors representing Hong Kong Chinese speakers and native English speakers. The HKCSE (prosodic) is unique in that it is transcribed for prosodic features using Brazil's (1985/1997) model of Discourse Intonation. A concordancing program called iConc was specifically developed for the corpus and allows quantitative analyses of intonational features (Cheng, Greaves, & Warren, 2008). The HKCSE has been utilized to document and quantify the intonation of *"yes/no"* questions, *wh-* questions, and declarative questions in service encounters; the intonation of disagreement sequences in business discourse; and the investigation of how participants give opinions in intercultural business discourse.

Over the past several years, corpora of aviation registers have been collected and analyzed, although the number of published corpus-based articles and research monographs in academia and the aviation industry is still very limited. Important collections such as The Air Traffic Control Communication Speech Corpus (ATCCSC) and the Air Traffic Control Simulation Speech Corpus (ATCOSIM) provide real-world spoken data that have been collected and utilized for teaching purposes. The ATCCSC is composed of a projected 140 hours of spoken interactions between pilots and ATC personnel (Šmídl & Ircing, 2014), while the ATCOSIM collects data from recordings of non-native English speakers during real-time air traffic control simulations (Hofbauer, Petrik, & Hering, 2008). A corpus-based study by Gabrielatos and Sarmento (2006) explored the frequency and distribution of central modals verbs in aviation using these types of corpora. Park's (2015) dissertation work at Iowa State University focused on a task-based needs analysis of eighty-one military ATCs on the required aviation English knowledge, skills, processes, target tasks, and task procedures, producing a corpus of interactions of virtual interactive aviation English tasks developed in Second Life. Although his primary interest was task-based performance assessment, his study also included related analyses of military aviation English training manuals and references and coded texts of Korean ATCs' stimulated recall and their actual task performance.

From Section 5.4 and Chapter 6, two sample case studies of CL and the spoken and written language of aviation are presented. The first study examines the lexico-syntactic features and readability of Aircraft Maintenance, Repair, and Overhaul (MRO) manuals with the traditional Flesch-Kincaid grade level formula and the *Coh-Metrix Easability Assessor* for reading texts. Five components from the corpus tool *Coh-Metrix* (Crossley & McNamara, 2009) are selected to measure the readability of MRO manuals, which include narrativity, syntactic simplicity, word concreteness, referential cohesion, and deep cohesion. These frequency-based features and distributions in MRO manuals are extracted and analyzed to describe this technical, written discourse in greater depth.

Chapter 6 compares the spoken, transactional discourse of global aviation with maritime interactions and other telephone-based speech registers, Switchboard discussions, and outsourced call center interactions. Switchboard texts came from transcribed recordings of two people discussing ideas and responses to a question or prompt (e.g., "What do you think about dress codes at work?"). Call center transactional talk parallels many of the typical interactions between pilots and ATCs. In the United States, third-party call centers specializing in training and hiring Indian and Filipino customer service representatives (or "agents") have increasingly staffed many US companies for very low salaries by current standards (Friginal, 2013a). This phenomenon has been possible because India, the Philippines, and other foreign nations offer tax breaks to outsourcers, allowing these companies to significantly reduce technical and operational expenses. Unlike in other cross-cultural business and workplace settings such as teleconferencing in multi-national company meetings or negotiations in international commerce and trade, business communications in outsourced call centers have clearly defined roles, protocols, power structures, and standards against which the satisfaction levels of customers during and after the transactions are often evaluated (Cowie & Murty, 2010; Lockwood, Forey, & Elias, 2009). Much like intercultural pilot and ATC exchanges, however, there are clear cultural and attitudinal factors that may directly influence how interlocutors make use of language (i.e., English) across various types of tasks. Here, CL approaches are utilized, with promising results.

5.4. Linguistic features of airplane maintenance manuals

As noted previously in this book English is regarded as the official language of written documents, especially technical documents produced by airplane manufacturers, typically in the form of instructional manuals and related troubleshooting and procedural texts. Airbus, Boeing, and even smaller manufacturers or distributors of airplane parts and equipment have their documents written and developed in English for users and stakeholders globally, assuming that those users have the necessary English reading and comprehension skills to accurately comprehend these documents. As one would expect, however, there is considerable variation in the English proficiency level across the entire population of international aviation personnel (Borowska, 2017; Ma, Drury, & Marin, 2010). For example, Brazilian and Chinese engineers and technicians reading the same text in English may not necessarily have the same level of reading and comprehension skills even as they address the same technical issue or concern.

The growing demand for technicians and maintenance workers worldwide vis-à-vis the default English texts for manuals and instructional or assembly texts has prompted efforts to simplify aviation written communication. The goal here is to attempt to reduce misunderstanding and enhance accurate and complete information exchange efficiency as manuals are written and presented to users. Written in Simplified Aviation English (SAE) or Simplified Technical English (STE), various Maintenance, Repair, and Overhaul (MRO) manuals produced by airplane manufacturers all have an inherent goal to reduce, for example, the reading difficulty of these texts through the use of limited vocabulary and short sentences (Borowska, 2017; STEMG, 2017; Werfelman, 2007). Lexical terms are often defined, redefined, and repeated;

long clauses are avoided; and narrative features are reduced to a minimum. SAE/STE writing follows what the industry refers to as "ASD-STE100 standards," developed to enhance the readability of manuals (Gabrielatos & Sarmiento, 2006; Pendić, Ćosović, Pendić, & Jakovljević, 2016; STEMG, 2017). Specifically, sentences in MROs are restricted to twenty to twenty-five words using allowable or approved dictionary words with easily comprehended forms and functions.

Note that these are still highly technical texts and not abridged to one that lay readers will easily understand. Below is an excerpt from an MRO manual describing the operation of an "engine bleed air distribution system."

Text Sample 5.1. MRO text description of "engine bleed air distribution system"

> The purpose of the engine bleed air distribution system is to direct high temperature compressed air from the engine bleeds to the air conditioning system, to the engine starting system, and to the thermal anti-icing system. The engine bleed system consists of ducts that interconnect the two engines, an isolation valve, and engine bleed valves to isolate each engine and deliver air as required.
>
> ...
>
> The right air conditioning system takes air from the No. 2 engine and the left air conditioning system from the No. 1 engine. (See Figure 2.) One isolation valve is provided for the isolation of the systems. The isolation valve is automatic in operation and electrically controlled.
>
> ...
>
> Normally the APU check valve prevents engine bleed pressure from backing into the APU bleed system. Should the check valve fail open, with the APU bleed valve open, engine bleed pressure could be detrimental to the APU.

In most international settings, technicians working for an airline company have ready access to English coaches and subject matter experts (SMEs) who can assist if there are instructions that are not clear. These technicians receive training and coaching in English, and the hiring process often includes interviews in English. Some may be required to provide TOEFL or IELTS test scores to qualify during the screening process.

5.4.1. Understanding SAE/STE in MRO manuals

English-only texts have been adopted globally, and only a few smaller airline companies make use of translated MROs, that is, English MROs translated into a local language. SAE/STEs, however, are still not fully standardized or consistently implemented universally in aviation written texts, and different companies may follow their own internal version of STE in their technical documents.

The Aerospace and Defense Industries Association of Europe (ASD), formerly known as the European Association of Aerospace Industries (AECMA), published the first set of STE documents in the late 1970s. As a controlled written language, STE focuses on minimizing the complexity of technical documents, using only a selective and approved vocabulary and simplified grammatical and syntactic structures. ASD-STE100 prescribed a principle of "one word-one meaning," with sentences written to be free of ambiguity and circumlocution (Werfelman, 2007). STE, however, does not regulate or standardize text formatting, including the use of visuals and images. In 2017, the ASD-STE100 was updated to its seventh edition, simplifying new technical terms and jargon and incorporating words that are originally from other fields such as computer science and industrial engineering. Examples of STE-approved vocabulary and sentence structures are provided in Table 5.2.

Table 5.2 Sample STE-approved vocabulary and sentences, adapted from Simplified Technical English Specification, ASD-STE100 (7th Ed.) (STEMG, 2017) Approved Vocabulary

Word (part of speech)	Approved meaning/ ALTERNATIVES	APPROVED EXAMPLE	Not approved
CATCH(v), CATCHES, CAUGHT, CAUGHT	To stop or prevent the movement of something	THE LUG ON THE PANEL OPENING CATCHES THE BOTTOM OF THE DRAWER.	
empty (v)	REMOVE (v)	REMOVE ALL FUEL FROM THE FUEL TANKS.	Empty the fuel tanks.
lack (n)	NOT SUFFICIENT	DAMAGE CAN OCCUR TO THE PUMP IF THERE IS NOT SUFFICIENT OIL IN THE RESERVOIR.	Damage can occur to the pump if there is a lack of oil in the reservoir.

Sample Sentences:

1) Making ambiguous sentences more specific

   Non-STE: *Different temperatures will change the cure time.*
   (This is an abstract sentence because it is not specific; it does not tell the reader how to decrease the cure time.)

   STE: *Increase the temperature to decrease the cure time.*
   (This sentence tells the reader exactly what to do to decrease the cure time.)

2) Replacing non-STE adjective "visible" with approved verb "see"

   Non-STE: *The oil level on the sight gauge must be visible during the test*
   STE: *Make sure that you can see the oil level on the sight gauge during the test.*

3) Replacing the unapproved verb "cycle" with approved noun "cycle"

   Non-STE: *Cycle the unit twice to remove air from the lines.*
   STE: *Operate the unit for two cycles to remove air from the lines.*

The STE examples above illustrate how approved words are restricted to be used only for the intended, particular part of speech, meaning, and form. Noun clusters are limited to no more than three words. For procedural or instructional writing, specifying steps and sequences, the active voice is preferred or recommended in most situations. In sentences, procedural and descriptive writing are clearly specified, with a recommended twenty-word per sentence length for procedural texts and twenty-five for descriptive texts. Focused instructional language is written strictly in the imperative (STEMG, 2017).

The obvious question here is, do STE standards and prescriptions really make a difference in the comprehension of technicians with limited English proficiency? In simplifying technical written language of MROs, have these documents actually been written to have enhanced readability and comprehension that many international technicians will understand relatively easily compared to other technical, non-STE texts?

Corpus of MRO manuals. The exploratory corpus of MRO manuals collected for this book has a total of 157 different technical documents from various aviation industry companies, with many of the manuals serving as actual worksheets or instructional texts for aircraft personnel. Some of the texts are used in training programs in settings such as universities or private flight schools in various locations worldwide. The texts are primarily public, with some available for purchase. The total number of words in the corpus is 5,673,223 (average words per document = 36,135). Short, instructional texts, stand-alone illustrations or models, or glossaries of technical terms are not included in the corpus of MRO manuals examined for analyses in this book.

Typical MRO manuals from airplane manufacturers feature chapters that, first, introduce sections and compartments of the aircraft, followed by specific descriptions and operating procedures. These manuals provide a detailed table of contents; a general introduction to the current edition or volume; a list of abbreviations and their meanings; and lists of tables, figures, and illustrations. There are clear differences in the manuals developed by major commercial airline manufacturers such as Airbus or Boeing, producing large-scale carriers and typically selling to major airlines, corporations, or countries in contrast to the relatively smaller companies such as Cessna or Bombardier Aerospace. Length is a clear indicator of the source of the manual; Boeing manuals, for example, are voluminous, with MROs of over 200,000 words on average per volume, not including figures and illustrations. As would be expected for the larger carriers, there are more chapters discussing the maintenance of specific compartments of the carrier, especially those that deal with fuel, air conditioning, and sections occupied by flight crew and passengers.

Most MRO manuals could be categorized into two textual parts (i.e., genres, in the area of genre analysis in applied linguistics): a *general* or descriptive section, which primarily introduces, defines, and describes in depth a specific part, item, or section of the carrier; and a *procedural* section, providing step-by-step instructions for various maintenance and troubleshooting needs. These sections may follow STE standards in length per sentence mentioned earlier, with descriptive sections typically being longer and perhaps also more grammatically complex than numbered and sequenced procedures as shown in the excerpts below (Text Sample 5.2):

Text Sample 5.2. Excerpt from an MRO: Descriptive vs. procedural texts

General: (Description)

> Before performing adjustment/test of the airstair and door, the control console, release handle and torsion bar assemblies must be properly installed and adjusted. The remaining major portions of the airstair must also be installed.

> Since the adjustments on the airstair are made without the use of the emergency extension mechanism, the return cam in the emergency extension mechanism must be removed to prevent accidental damage to the mechanism. The emergency extension system is adjusted after the airstair and door are installed and adjusted. Refer to 52–14-21, Emergency Extension Mechanism—Adjustment/Test.

Adjust Entry Door: (Procedures)

1. Remove door from airstair and remove airstair. Refer to Aft Entry Door and Airstair-Removal/Installation.
2. Remove top, side and bottom door lining panels if installed. (Refer to 52–14-131, Aft Airstair Door Lining.)
3. Remove door pressure seal.
4. Remove all bolts in door cam fitting (2) except the upper inboard bolt on each fitting. This bolt must be loosened.
5. Remove locksprings (10) and all stop pins except at lower forward, lower aft and upper forward fitting locations. Adjust three remaining stop pins to approximately 0.40-inch protrusion.

5.4.2. Vocabulary features of MRO manuals

This section presents various results of a simple frequency-based analysis of the MRO manuals corpus using basic concordancing or key-word in context (KWIC) software. One of the most basic types of analysis that corpora offer researchers is the ability to determine the frequency of linguistic items; for example, words are the most frequently used words in a particular register or genre. Although many function words like determiners (e.g., *the, a*) and prepositions (e.g., *to, of*) consistently top frequency lists across varieties and English, there is great diversity in distributions, especially when it comes to lexical words (i.e., "ordinary words" or meaning-carrying words) when one looks more closely (Friginal & Hardy, 2014a). In aviation texts, frequency is important in both the description of the broader discourse and also in determining what to focus on when developing materials for teaching.

Most frequent words in MRO manuals. From the MRO manuals corpus, Table 5.3 lists the top 200 most frequent words, with lexical words highlighted. Most frequent nouns include *engine, air, bleed, valve,* and *name,* while the most frequent verbs are *remove, check, install, mount,* and *access* all in the top 50. Clearly, it will have to be confirmed in concordance lines to check the actual POS category of a word (e.g., *access* as a verb or a noun) and its actual meaning or use. High-frequency words unique to aviation texts can be found on the list below (e.g., *pneumatic, manifold, cotter,* or *inlet*).

Most frequent collocates. Firth's (1957) study of collocations explored discrete linguistic elements (or words) and their common components, appearing before or after each single word to form a consistent pattern in the discourse. Collocational patterns from an aviation corpus add another layer of meaning in interpreting discourse characteristics that may be unique to this domain. Collocations can be extracted using more objective measurements

Table 5.3 Top 200 most frequent words in the MRO manuals corpus

Rank	Word	Rank	Word	Rank	Word
1	the	68	exchanger	135	hydraulic
2	and	69	manifold	136	latches
3	to	70	pack	137	nuts
4	of	71	stage	138	point
5	engine	72	bolt	139	remove
6	air	73	provided	140	require
7	is	74	ring	141	thermal
8	bleed	75	supply	142	all
9	valve	76	clamps	143	approximately
10	on	77	equipment	144	bay
11	name	78	left	145	closure
12	proper	79	off	146	consists
13	duct	80	pin	147	doors
14	a	81	through	148	relief
15	in	82	an	149	section
16	forward	83	chapter	150	door
17	remove	84	cotter	151	fitting
18	from	85	five	152	light
19	for	86	inch	153	necessary
20	system	87	joint	154	ninety
21	check	88	midfairing	155	shutoff
22	pressure	89	which	156	steel
23	install	90	circuit	157	threads
24	are	91	outboard	158	three
25	with	92	right	159	transmitter
26	that	93	wear	160	zero
27	one	94	downstream	161	after
28	ducts	95	removal	162	airplanes
29	switch	96	side	163	any
30	by	97	this	164	bonding
31	aft	98	between	165	closed
32	or	99	bolts	166	controlled
33	valves	100	connector	167	disconnect
34	at	101	control	168	engines

(Continued)

Rank	Word	Rank	Word	Rank	Word
35	mount	102	may	169	ground
36	two	103	overheat	170	has
37	leakage	104	washers	171	inlet
38	be	105	fifty	172	other
39	hundred	106	seventy	173	rate
40	panel	107	end	174	thirteenth
41	as	108	flow	175	used
42	test	109	power	176	bearing
43	pneumatic	110	twenty	177	distribution
44	position	111	airplane	178	fire
45	not	112	area	179	fluid
46	isolation	113	attached	180	gain
47	access	114	drain	181	indicator
48	if	115	fixed	182	inspection
49	nut	116	inboard	183	insulation
50	wing	117	joints	184	it
51	installed	118	located	185	link
52	four	119	support	186	main
53	heat	120	a	187	plug
54	installation	121	both	188	port
55	open	122	clean	189	prepare
56	each	123	line	190	self
57	cowl	124	pounds	191	source
58	per	125	securing	192	tests
59	temperature	126	switches	193	then
60	overhead	127	thrust	194	thermostat
61	conditioning	128	will	195	torque
62	fairing	129	APU	196	turbofan
63	no	130	ball	197	anti
64	precooler	131	breaker	198	applicable
65	close	132	eighth	199	apply
66	degrees	133	flange	200	assembly
67	electrical	134	high		

from statistical results obtained from running common KWIC software like *AntConc* (2018). Prediction models of what might follow or precede a word, a noun, or a verb can be measured based on their expected frequencies. And, as the common argument here, these may be utilized to teach the form and structure of high-frequency words in aviation. Tables 5.4 and 5.5 list the top collocates of highly-frequent noun (*engine*) and verb (*install*) from the MRO manuals corpus.

Table 5.4 Top collocates of *engine* from the MRO manuals corpus

Rank	Freq.	1-Left	1-Right	Collocates
1	290	0	290	bleed
2	266	0	66	mount
3	262	262	0	the
4	262	262	0	forward
5	242	240	2	the
6	220	4	216	to
7	220	216	4	remove
8	220	2	218	is
9	218	218	0	each
10	218	0	218	cowl
11	218	6	212	and
12	216	0	216	no
13	214	106	108	one
14	214	214	0	of
15	212	110	102	install
16	210	0	210	per
17	210	210	0	from
18	210	108	102	for
19	208	0	208	starting
20	208	208	0	stage
21	208	208	0	right
22	208	208	0	respective
23	208	0	208	bleeds
24	208	208	0	aft
25	206	0	206	vibration

(Continued)

Rank	Freq.	1-Left	1-Right	Collocates
26	206	104	102	two
27	206	0	206	start
28	206	206	0	name
29	206	206	0	left
30	204	0	204	supply
31	204	204	0	prevents
32	204	204	0	position
33	204	0	204	ports
34	204	0	204	fan
35	204	0	204	equipment
36	204	0	204	cowls
37	204	0	204	compressor
38	204	204	0	close
39	204	204	0	by
40	204	204	0	an
41	202	202	0	whenever
42	202	0	202	when
43	202	202	0	valve
44	202	202	0	uses
45	202	0	202	this
46	202	202	0	test
47	202	0	202	support
49	202	202	0	structural
50	202	0	202	shutdown

Table 5.5 Top collocates of *install* from the MRO manuals corpus

Rank	Freq	1-Left	1-Right	Collocates
1	248	0	248	the
2	212	0	212	closure
3	210	102	108	engine
4	210	108	102	duct
5	210	106	104	and
6	208	208	0	mount
7	208	208	0	installation
8	206	206	0	pounds

Rank	Freq	1-Left	1-Right	Collocates
9	206	206	0	pin
10	206	206	0	fairing
11	206	206	0	ducts
12	206	0	20	bonding
13	206	102	104	aft
14	204	102	102	valve
15	204	0	204	two
16	204	204	0	then
17	204	0	204	proper
18	204	0	204	nuts
19	204	204	0	not
20	204	204	0	found
21	204	0	204	forward
22	204	204	0	exchanger
23	204	102	102	clamps
24	204	102	102	bolts
25	204	0	204	access
26	202	0	202	wear
27	202	202	0	transmitter
28	202	202	0	torque
29	202	202	0	system
30	202	0	202	switch
31	202	202	0	strap
32	202	0	202	split
33	202	202	0	shims
34	202	202	0	section
35	202	0	202	seal
36	202	0	202	screws
37	202	202	0	ring
38	202	202	0	require
39	202	0	202	precooler
40	202	202	0	position
41	202	0	202	pneumatic
42	202	202	0	places
43	202	202	0	panels
44	202	202	0	on
45	202	202	0	midfairing

(Continued)

Rank	Freq	1-Left	1-Right	Collocates
46	202	202	0	material
47	202	202	0	loosely
48	202	0	202	lockring
49	202	202	0	limits
50	202	202	0	latches

Most frequent multi-word units (MWU) in MRO manuals. Multi-word units (MWUs) are expanded collocations frequently occurring as linear strings, similar to prefabricated "chunks" of language. Research on MWUs has shown that discourse communities (and speech communities) use such chunks differently, and as expected, aviation MWUs are more technically defined and very discreet compared to other groups of texts. There are various ways to explore this construct of formulaic language using corpora; three of the commonly used approaches are n-grams, lexical bundles, and p-frames. **N-grams** are the most common MWUs, typically measured through bi-grams (two words in order, essentially collocates), tri-grams (three words), four-grams, etc. **Lexical bundles** are a type of N-grams, but there are more limitations or specifications as to how they are extracted or categorized (or also defined). Typically, lexical bundles consist of at least three words (tri-grams) that occur frequently across a corpus of at least 1 million words. This is determined by a count per 1 million words. The frequency, however, can be determined by the researcher. Another important criterion for labeling MWUs as lexical bundles is that they surface in at least five different texts in the corpus. This is necessary to avoid any idiosyncratic language usages (Biber, Conrad, & Cortes, 2004). And finally, **p-frames** are frequent, patterned constructions of phraseological units that allow for variability in one position. For example, a p-frame found by Römer (2010) from an academic written corpus is *"it would be _____ to,"* with *interesting, useful, nice*, and *better* as the most frequent words in the slot, accounting for 77 percent of all the variants in the corpus (Friginal & Hardy, 2014a). Tables 5.6 and 5.7 provide the most common 4-grams, 5-grams, and 6-grams of the MRO manuals corpus.

Keyword analysis of the MRO manuals corpus. Keyword analysis requires a statistical procedure that identifies significant differences in the distribution of words from a particular corpus compared to another (Baker, 2006; Barbieri, 2008; Scott, 1997). The frequency of a word, as present in a general frequency list, is typically compared using statistical tests against distributions from a reference corpus to extract the actual, measurable range of difference. Scott (1997) defines a keyword as "a word which occurs with unusual frequency in a given text" (p. 236). This "unusual frequency" is based on the likelihood

Table 5.6 Most frequent 4-grams of the MRO manuals corpus (appearing at least 40 times)

No.	F	4-grams	No.	F	4-grams
1	76	the forward overhead panel	51	42	nut to a value
2	74	on the forward overhead	52	42	of grease and torque
3	63	COTTER PIN INSTALLED ON	53	42	of the bleed air
4	60	four hundred and ninety	54	42	power if no longer
5	57	hundred and ninety degrees	55	42	Power Plant Removal Installation
6	57	THE COTTER PIN INSTALLED	56	42	Remove electrical power if
7	56	PIN INSTALLED ON THE	57	42	Restore Airplane to Normal
8	50	air conditioning equipment bay	58	42	sure the bolt threads
9	50	four hundred and fifty	59	42	the bolt threads and
10	50	ISOLATION VALVE switch to	60	42	the engine bleed air
11	50	outboard forward engine mount	61	42	the four hundred and
12	50	switch on the forward	62	42	the ISOLATION VALVE switch
13	50	to gain access to	63	42	thirteenth stage modulating and
14	48	engine mount to the	64	42	threads are free of
15	48	forward engine mount to	65	42	to a value of
16	48	is attached to the	66	42	WITH THE COTTER PIN
17	48	mount to the wing	67	42	WITHOUT THE COTTER PIN
18	48	the outboard forward engine	68	41	Chapter seventy-one Power
19	46	conditioning equipment bay doors	69	41	one Power Plant Removal
20	46	forward overhead panel to	70	41	seventy-one Power Plant
21	46	Install the bolt washers	71	40	air to air heat
22	46	of the engine bleed	72	40	and bolt securing the
23	46	the bolt washers and	73	40	and ninety degrees overheat
24	46	the forward engine mount	74	40	and self-locking nut
25	44	bleed air distribution system	75	40	and torque the nut
26	44	engine bleed air distribution	76	40	attached to the forward

(Continued)

No.	F	4-grams	No.	F	4-grams
27	44	engine mount as follows	77	40	bolt securing the forward
28	44	forward engine mount as	78	40	bolt washers and nut
29	44	if no longer require	79	40	Close air conditioning equipment
30	44	modulating and shutoff valve	80	40	compressed air from the
31	44	OUTBOARD FORWARD ENGINE MOUNT	81	40	engine bleed air valve
32	44	per Chapter seventy-one	82	40	exceeds four hundred and
33	44	proper name bleed air	83	40	fire resistant hydraulic fluid
34	44	THE OUTBOARD FORWARD ENGINE	84	40	INBOARD FORWARD ENGINE MOUNT
35	43	INSTALLED ON THE OUTBOARD	85	40	INSTALLED ON THE INBOARD
36	43	of the heat exchanger	86	40	ISOLATION VALVE switch on
37	43	ON THE OUTBOARD FORWARD	87	40	leakage rate does not
38	42	AIRPLANES WITH THE COTTER	88	40	mounting boss on duct
39	42	AIRPLANES WITHOUT THE COTTER	89	40	nut washers and bolt
40	42	are free of grease	90	40	ON THE INBOARD FORWARD
41	42	electrical power if no	91	40	per Chapter twenty-seven
42	42	forward overhead panel position	92	40	pneumatic duct pan assembly
43	42	free of grease and	93	40	proper name bleed valve
44	42	grease and torque the	94	40	proper name clamps and
45	42	hundred and fifty degrees	95	40	rate does not exceed
46	42	inboard forward engine mount	96	40	securing the forward engine
47	42	is located in the	97	40	side of each engine
48	42	is provided for the	98	40	stage modulating and shutoff
49	42	Make sure the bolt	99	40	the air conditioning system
50	42	nut threads are free	100	40	the engine to wing

Table 5.7 Most frequent 5-grams and 6-grams of the MRO manuals corpus

Top 5-grams (appearing at least 20 times)

1	on the forward overhead panel
2	four hundred and ninety degrees
3	THE COTTER PIN INSTALLED ON
4	COTTER PIN INSTALLED ON THE
5	switch on the forward overhead
6	engine mount to the wing
7	forward engine mount to the
8	the outboard forward engine mount
9	air conditioning equipment bay doors
10	Install the bolt washers and
11	the forward overhead panel to
12	engine bleed air distribution system
13	forward engine mount as follows
14	THE OUTBOARD FORWARD ENGINE MOUNT
15	INSTALLED ON THE OUTBOARD FORWARD
16	ON THE OUTBOARD FORWARD ENGINE
17	PIN INSTALLED ON THE OUTBOARD
18	AIRPLANES WITH THE COTTER PIN
19	AIRPLANES WITHOUT THE COTTER PIN
20	are free of grease and
21	electrical power if no longer
22	four hundred and fifty degrees
23	free of grease and torque
24	Make sure the bolt threads
25	nut threads are free of
26	nut to a value of
27	of grease and torque the
28	power if no longer require
29	Remove electrical power if no
30	sure the bolt threads and
31	threads are free of grease
32	WITH THE COTTER PIN INSTALLED
33	WITHOUT THE COTTER PIN INSTALLED
34	Chapter seventy-one Power Plant

(Continued)

35	one Power Plant Removal Installation
36	seventy-one Power Plant Removal
37	air to air heat exchanger
38	and bolt securing the forward
39	and torque the nut to
40	bolt securing the forward engine
41	Close air conditioning equipment bay
42	grease and torque the nut
43	hundred and ninety degrees overheat
44	INSTALLED ON THE INBOARD FORWARD
45	ISOLATION VALVE switch on the
46	ISOLATION VALVE switch to OPEN
47	leakage rate does not exceed
48	nut washers and bolt securing
49	of the engine bleed air
50	ON THE INBOARD FORWARD ENGINE

Top 6-grams (appearing at least 15 times)

1	THE COTTER PIN INSTALLED ON THE
2	switch on the forward overhead panel
3	forward engine mount to the wing
4	on the forward overhead panel to
5	COTTER PIN INSTALLED ON THE OUTBOARD
6	INSTALLED ON THE OUTBOARD FORWARD ENGINE
7	ON THE OUTBOARD FORWARD ENGINE MOUNT
8	PIN INSTALLED ON THE OUTBOARD FORWARD
9	AIRPLANES WITH THE COTTER PIN INSTALLED
10	AIRPLANES WITHOUT THE COTTER PIN INSTALLED
11	are free of grease and torque
12	electrical power if no longer require
13	free of grease and torque the
14	Make sure the bolt threads and
15	nut threads are free of grease
16	Remove electrical power if no longer
17	threads are free of grease and
18	WITH THE COTTER PIN INSTALLED ON
19	WITHOUT THE COTTER PIN INSTALLED ON

(Continued)

20	Chapter seventy-one Power Plant Removal
21	seventy-one Power Plant Removal Installation
22	and bolt securing the forward engine
23	and torque the nut to a
24	bolt securing the forward engine mount
25	Close air conditioning equipment bay doors
26	COTTER PIN INSTALLED ON THE INBOARD
27	four hundred and ninety degrees overheat
28	grease and torque the nut to
29	Install the bolt washers and nut
30	INSTALLED ON THE INBOARD FORWARD ENGINE

of occurrence of the word in a target text from cross-tabulation. To identify keywords in a corpus, it is necessary to compare data from the corpus with a reference (or target) corpus that is logically representing similar linguistic characteristics and qualities (Friginal, 2009). Data from keyword analysis can be obtained from *WordSmith Tools* or *AntConc*.

Corpus-based studies of keywords have covered topics such as the utterances of older versus younger speakers in informal conversation (Barbieri, 2008) or academic writing in a particular discipline such as biology compared to other fields of study (e.g., history, political science, or English) (Römer & Wulff, 2010). Barbieri's (2008) keyword analysis of age-based variation explored parallel corpora grouped according to speakers' ages ("Younger Corpus" vs. "Older Corpus"). She found that younger speakers used unusually frequent slang and swear words, stance makers, and "emotional involvement" markers (e.g., intensifiers, discourse markers, personal pronouns, and attitudinal adjectives) compared to older speakers. Römer and Wulff (2010), using texts from the Michigan Corpus of Upper-Level Student Papers (MICUSP), reported that the words *species, gene(s), plague, cells*, and *protein* received highest keyness values for writing in biology, demonstrating that a keyword list can be a useful tool in the disciplinary writing classroom because it highlights items (e.g., specific words or characteristics) that are important in a certain discipline that students may need to know. Words that occur comparatively more often in other disciplines compared to biology (i.e., negative keywords) are *perceive, fields, organizational, metaphor*, and *government*.

To extract keywords from the MRO manuals corpus in this section, a collection of technical and professional written texts from fields such as engineering, industrial technology, and computer science was used as reference corpus. This reference corpus has a total of 477 texts with 6,455,744 words. The MRO manuals corpus was "cleaned" to remove, annotations and markups, and figures and tables, and both corpora were converted to all lower case to make sure that the keyword program properly captured and computed keywords and keyness values. Keyword comparisons provide an interesting picture of the unique features of MRO manuals as a written register of English and aviation language compared to other technical writing texts. Table 5.8 shows the 200 most common keywords of MRO manuals compared to technical written texts in other fields.

Table 5.8 Top keywords based on keyness values of the MRO manuals corpus

No.	Keyness Value	Keyword	No.	Keyness Value	Keyword
1	1024.999	engine	101	64.684	latches
2	820.994	bleed	102	64.684	nuts
3	788.928	air	103	64.684	relief
4	773.987	valve	104	64.684	transmitter
5	671.723	name	105	63.464	provided
6	671.723	proper	106	59.709	doors
7	537.378	duct	107	59.709	plug
8	520.346	remove	108	59.709	repair
9	492.597	forward	109	59.709	shutoff
10	422.936	check	110	59.709	steel
11	383.205	pressure	111	59.709	thermostat
12	381.631	install	112	59.709	tighten
13	333.974	system	113	56.57	drain
14	323.422	switch	114	54.878	control
15	288.592	ducts	115	54.733	door
16	282.698	valves	116	54.733	engines
17	263.713	aft	117	54.733	ninety
18	263.713	mount	118	54.733	threads
19	238.835	leakage	119	54.733	titanium
20	238.835	pneumatic	120	54.733	torque
21	223.908	panel	121	51.992	apply
22	210.886	hundred	122	51.992	bay
23	191.343	position	123	51.992	closure
24	188.932	access	124	51.992	consists

CORPUS-BASED ANALYSIS OF AVIATION DISCOURSE

No.	Keyness Value	Keyword	No.	Keyness Value	Keyword
25	188.932	wing	125	51.992	refer
26	179.126	nut	126	51.988	normal
27	169.717	is	127	49.757	bonding
28	165.12	isolation	128	49.757	inspection
29	164.319	installed	129	49.757	thirteenth
30	164.199	fairing	130	47.664	end
31	164.199	precooler	131	47.434	require
32	159.223	installation	132	47.186	fitting
33	149.272	conditioning	133	47.186	zero
34	149.272	cowl	134	46.276	fixed
35	149.272	pack	135	45.469	per
36	144.296	overhead	136	44.782	anti
37	143.234	the	137	44.782	bearing
38	140.811	heat	138	44.782	grease
39	129.369	degrees	139	44.782	insulation
40	129.369	electrical	140	44.782	prepare
41	129.369	exchanger	141	44.782	replace
42	129.369	outboard	142	44.782	trip
43	124.393	overheat	143	44.782	turbofan
44	124.393	supply	144	42.912	section
45	116.499	test	145	42.397	fluid
46	115.229	ring	146	40.407	through
47	114.442	bolt	147	39.806	assembly
48	114.442	clamps	148	39.806	bushings
49	114.442	midfairing	149	39.806	cooling
50	114.442	wear	150	39.806	firewall
51	110.334	removal	151	39.806	flap
52	109.466	pin	152	39.806	forty
53	105.443	joint	153	39.806	locking
54	104.49	chapter	154	39.806	modulating
55	104.49	cotter	155	39.806	movable
56	104.49	inboard	156	39.806	nose
57	104.49	inch	157	39.806	opening
58	104.49	power	158	39.806	panels
59	103.376	manifold	159	39.806	plate

(Continued)

No.	Keyness Value	Keyword	No.	Keyness Value	Keyword
60	99.514	circuit	160	39.806	transmitters
61	95.671	downstream	161	37.626	manual
62	95.671	right	162	37.626	port
63	93.934	stage	163	36.962	line
64	93.91	equipment	164	36.669	two
65	92.552	close	165	34.83	bleeds
66	89.563	bolts	166	34.83	boss
67	89.563	washers	167	34.83	fasteners
68	88.866	on	168	34.83	fittings
69	88.073	to	169	34.83	inches
70	86.678	one	170	34.83	plates
71	85.918	connector	171	34.83	removing
72	84.65	off	172	34.83	restore
73	81.352	open	173	33.138	link
74	80.018	side	174	33.138	operation
75	79.612	joints	175	32.879	applicable
76	79.612	thrust	176	32.879	mounting
77	79.612	twenty	177	32.258	light
78	78.461	temperature	178	32.258	necessary
79	76.185	fifty	179	30.851	inlet
80	75.114	flow	180	30.183	no
81	74.636	airplane	181	29.854	adjust
82	74.636	airplanes	182	29.854	bracket
83	74.636	attached	183	29.854	cart
84	74.636	ball	184	29.854	cracks
85	74.636	switches	185	29.854	discharge
86	70.499	seventy	186	29.854	flaps
87	69.66	clean	187	29.854	flight
88	69.66	hydraulic	188	29.854	icing
89	69.66	pounds	189	29.854	latch
90	69.66	securing	190	29.854	leak
91	69.66	thermal	191	29.854	lockwire
92	68.663	if	192	29.854	packing
93	66.481	gain	193	29.854	pan
94	66.003	left	194	29.854	pressurize
95	64.684	APU	195	29.854	pressurized

No.	Keyness Value	Keyword	No.	Keyness Value	Keyword
96	64.684	breaker	196	29.854	retainer
97	64.684	connect	197	29.854	sensing
98	64.684	disconnect	198	29.854	shall
99	64.684	eighth	199	29.854	sure
100	64.684	flange	200	29.854	troubleshooting

Clearly, the keywords in MRO manuals compared to related technical written texts are primarily nouns and proper nouns, again repeating common words captured in the general word list and MWUs presented in the previous sections. *Engine, bleed, air*, and *valve* stood out as top keywords, repeated with high-frequency throughout the manuals in sentences such as: "*Bleed air produced by gas turbine engines is compressed air that is taken from the compressor stage of those engines, which is upstream of the fuel-burning sections.*" Most of the top 200 keywords are, for the most part, very common words, with only a few of them carrying a very specific meaning in aviation compared to other disciplines or popular/lay usage (e.g., *bleed, pneumatic, lockwire*). These common words identified as keywords in aviation texts support efforts to simplify and standardize vocabulary in aviation through SAE/STE conventions.

5.5. The readability of MRO manuals

According to Ma, Drury, and Marin (2010), as mentioned in Chapter 4, the traditional reading grade level (in English) of US-based technicians in the aviation industry is around 14. In contrast, for technicians whose first language is not English in various global locations, the average reading grade level in English is only 5. Ma, Drury, and Marin's data also show that English proficiency levels are significantly lower in Asian sites. Specifically, in the Asia-Pacific region, with a vast number of technicians coming from China, only 40 percent of maintenance workers are able to accurately and effectively follow instructions from MROs without coaching or support. In addition, 15 percent of these technicians have an average English grade level of close to 1. The Asia-Pacific is the fastest-growing region for airplane MRO operations, especially for major companies, including FedEx and several US-based carriers.

Investigations into the linguistic features of MRO manuals reveal not only their readability features but also how simplification standards can influence the writing of technical manuals and how some aspects of the standards could be modified to enhance readability. The revision of simplification standards for aircraft manuals could also shed light on text simplification in other fields. Although STE has prescribed limited vocabulary, there are studies that show that only 3 percent of words are specific to aviation. Note that currently, the STE is utilized not only in aviation but also widely applied in other related fields. Thus, research focusing on improving the comprehensibility of aviation MRO manuals will also be useful to other types of technical writing in similar fields (Borowska, 2017).

Kaji's seminal study (1999) introduced the development of human-oriented and machine translation (MT)-oriented controlled language. Similar to the structure of the first practical controlled language, Caterpillar Fundamental English (CFE), which was designed for service manual writing for non-native operators, STE is also a human-oriented controlled language. Companies such as Caterpillar Inc. have already been working on replacing CFE with Caterpillar Technical English (CTE) and a machine translation system, anticipating future revision of human-oriented STE into a more MT-oriented language which would enable the development of STE machine translation and automatic STE checkers. STE supporting software, such as the Boeing Online Simplified English Checker (www.boeing.com/phantom/sechecker/se.html), has been developed to check whether or not the writing is consistent with ASD-STE100 standards.

Similar to previous studies investigating textual features of written texts (e.g., Crossley, Greenfield, & McNamara, 2008; Ismail, Yusof, & Yunus, 2016; McNamara et al., 2011, 2012; Ozuru, Dempsey, & McNamara, 2009), this section makes use of the Flesch Kincaid Grade Level index and five additional indices from the *Coh-Metrix Easability Assessor* (Crossley, Allen, & McNamara, 2011) to explore the readability of MRO manuals. The traditional Flesch Kincaid formula relates text readability to US grade school levels, which range from 0 to 12, with 0 representing the easiest and 12 being the most difficult reading level. Five easability indices, including narrativity, word concreteness, syntactic simplicity, referential cohesion, and deep cohesion, are also used to describe the readability of MRO manuals, with a scale of high (85–100 percent, 100 percent being the simplest), average (26–74 percent), and low (0–25 percent). The minimum, maximum, mean, and standard deviation of the five indices are shown in the results below. In summary, presented below are: (1) the MRO manuals' overall readability score and resulting readability indices from *Coh-Metrix Easability Assessor*, and Flesch-Kincaid grade level; (2) the readability scores of descriptive vs. procedural texts; and (3) summary of results and pedagogical applications.

5.5.1. Brief overview of readability measures

Over the years, there have been numerous models and formulas developed to estimate text readability for intended readers. Before the inclusion of fields such as computer science and cognitive psychology on readability tests, most readability algorithms measured only surface-level features such as word and sentence length (Benjamin, 2011; Crossley, Allen, & McNamara, 2011). Two of the traditional measures, the Flesch-Kincaid grade level and the Flesch reading ease (Flesch, 1948; Kincaid, Fishburne, Rogers, & Chissom, 1975), were first developed for the US Army to assess the difficulty of technical documents from a range of sources. These measures were then widely used in other disciplines, specific texts, and reading situations (Greenfield, 2004) for improving the readability of text genres such as insurance policies (McClure, 1987), controlling the difficulty of second language (L2) reading materials (Madhumathi & Ghosh, 2016), and analyzing students' writing (Doolan, 2014). Clearly, however, the accuracy and evaluative import of these traditional algorithms have always been explored in reading research as they do not incorporate, for example, cognitive processes involved in reading comprehension and textual factors such as syntactic complexity and idea organization (Carrell, 1987; Crossley, Allen, & McNamara, 2011), among others. Thus, more comprehensive readability measures that take reader and textual features into thorough consideration have been developed to address the limitations of traditional measures.

Graesser and McNamara (2011) proposed a multi-level framework of indices to investigate discourse or text comprehension. They focused on a five-level framework that includes surface code, textbase, situation model, genre and rhetorical structure, and pragmatic communication. This framework paved the way for the creation of indices that have been included in the suite of tools for the software *Coh-Metrix* (Crossley & McNamara, 2009). On the word level, *Coh-Metrix* evaluates word frequency with CELEX database, calculates polysemy and hypersemy values through WordNet, and identifies parts of speech with the tags from the Penn Treebank. With sentences, *Coh-Metrix* assesses sentence syntax through modifiers per noun phrase, words before the main verb of the main clause. Four indices are selected to describe textbase, including coreference, pronoun anaphors, connectives, and discourse markers that link clauses and sentences, and lexical diversity. The situation model can be classified into five dimensions, including causation, intentionality, time, space, and protagonists. It is then argued that lacking one or more dimensions will lead to comprehension breakdowns. In detail, the situation model is measured through causality and intentionality, temporality, spatiality, latent semantic analysis, and given information. In terms of genre and rhetorical composition, genre and topic "sentencehood" are also assessed by *Coh-Metrix* (Graesser & McNamara, 2011).

Crossley and his colleagues (2008) used three indices to measure text cohesion and readability: (1) a lexical index that analyzes word frequency through CELEX frequency scores (i.e., frequent words are processed faster and understood better than less frequent words); (2) a syntactic index that measures syntax similarity (i.e., the consistency and resemblance of parallel syntactic constructions lead to easier text decoding); and (3) lexical coreferentiality, which measures content word overlap (i.e., vocabulary overlaps could facilitate text comprehension). Their results show that linguistic features closely related to cognitive processes reflect the decoding, parsing, and meaning construction in reading, which can better describe text readability than can surface-level measures. They argue that Text Readability and Easability Indices from *Coh-Metrix* can accurately assess the linguistic characteristics of texts that contribute to comprehension difficulty. Although specific measures vary across contexts, their indices successfully produce encouraging descriptors in six general categories, including descriptive indices, words and sentences, referential cohesion, lexical diversity, connectives, and situation model.

5.5.2. *Exploratory results*

Flesch-Kincaid reading grade levels. A comparison of Flesch-Kincaid reading grade levels of all MRO manuals and specifically all descriptive and procedural texts is provided in Figure 5.1. MRO manuals, written in simplified English, are relatively very easy to read and comprehend at below 8th grade level. Note, again, that the average reading grade level for English-speaking technicians is 14 (Ma, Drury, & Marin, 2010). However, this result is, therefore, alarming for technicians outside of the United States and who are non-native speakers of English where the average reading grade level is only 5. This differential of close to 3 grade levels poses major challenges in technicians' comprehension of already simplified language. Aside from China, other countries with technicians tested for English reading grade level will also have clear challenges as reported by Ma, Drury, and Marin (2010). The average reading level and years of learning English of technicians from other target countries are:

(1) Mexico, reading level = 5.6; ave. years of learning English = 3.5

(2) Colombia, reading level = 4.9; ave. years of learning English = 3.3

(3) Argentina, reading level = 4.8; ave. years of learning English = 4.4

Technicians from Puerto Rico had the highest scores in the group, with reading level = 10.0; ave. years of learning English = 9.5.

Manuals (or sections of manuals) focusing on contextual descriptions of airplane or engine parts, that is, descriptive texts are harder to read and comprehend than procedural texts (10.62 vs. 6.98). Procedural texts have shorter, numbered items and they follow simple parallel structures. Descriptive texts, as shown below, are generally similar to academic or professional writing common in most manuals and textbooks.

Text Samples 5.3. Comparison of descriptive and procedural texts

Descriptive Text:
The right air conditioning system takes air from the No. 2 engine and the left air conditioning system takes air from the No. 1 engine. This separation of bleed air is accomplished by the isolation valve. The isolation valve remains closed unless a cross-feed of bleed air is required. The isolation valve is motor-driven and electrically controlled by a three-position switch on the forward overhead panel.

Procedure Text:
Restore airplane to normal.

(a) Release pressure and disconnect air supply.
(b) Remove duct closure plates.
(c) Connect warm air supply ducts to check valves and water separators to their outlet ducts.
(d) Install APU relief valve and APU bleed air valve.
(e) Install turbofan valve electrical connectors and purge valve electrical connectors

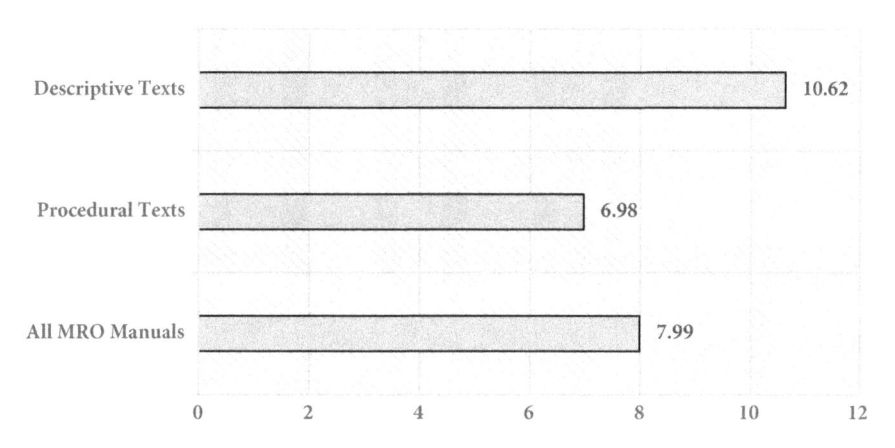

FIGURE 5.1 *Flesch-Kincaid reading grade levels of all MRO manuals vs. descriptive and procedural texts.*

Text readability and easability scores. Figure 5.2. compares the text readability and easability scores of all MRO manuals vs. descriptive and procedural texts, specifically from the following indices: (1) narrativity, (2) syntactic simplicity, (3) word concreteness, (4) referential cohesion, and (5) deep cohesion.

MRO manuals have an "average" referential cohesion and deep cohesion scores (below 85 percent), with deep cohesion significantly lower. These scores suggest that the presentation of concepts in MRO texts internally overlaps, but with minimal use of connectives and linking adverbials. Thus, readers may have to employ additional cognitive effort in processing the relationships between ideas and concepts in MROs, which, in turn, requires focused reading strategies and also for readers to be more familiar with written conventions of MROs. For sentence structure, the syntactic simplicity index showed a very "high" score of 97.83 percent, reflecting fewer words in a sentence (i.e., shorter sentences) and the repetition of similar structures. Employing careful control of sentence length and repetitive verb use, as well as consistent tense structure, MRO manuals are consequently syntactically simple according to this index, thereby achieving a major goal of STE.

Regarding word concreteness, as an informative or informational written genre, MROs, especially descriptive texts, utilize an "average" frequency of

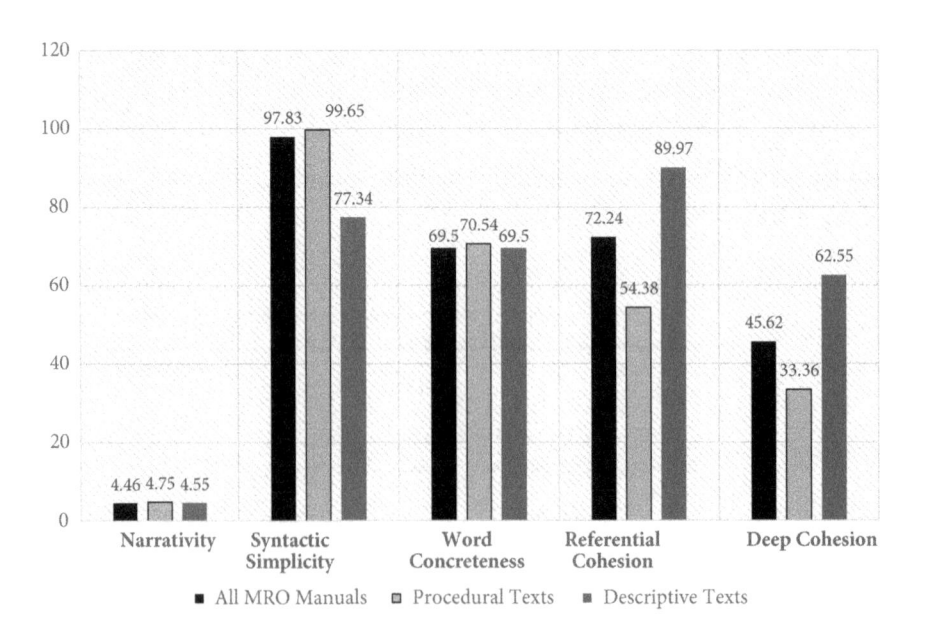

FIGURE 5.2 *Text readability and easability scores of all MRO manuals vs. descriptive and procedural texts, all in percentages. [For detailed descriptions of these indices, please see* Graesser and McNamara (2011) *and* Crossley, Greenfield, and McNamara (2008).*]*

concrete words. The ASD-STE-approved dictionary supports this result, with words specifically defined and repetitively used in context (e.g., *bleed, latch, outboard*). A low narrativity score (4.46 percent) denotes a formal writing style that requires more familiarity with technical terms and in this particular setting, aviation language, with very limited occurrences of less familiar words. As expected, narrativity scores are all very low across text groups in the MRO manuals corpus.

5.5.3. Pedagogical implications and recommendations for ASD-STE100 standards editors and manual writers

STE simplification strategies appear to significantly enhance MROs' text readability based on traditional readability models and especially for native English-speaking technicians. However, the clear implication of the comparative results in this section is that global aviation technicians, who are exponentially increasing in number, are not primarily supported by these existing simplification strategies. There still is a wide gap between the computed readability scores of MRO texts and the reading abilities of many global technicians which STE is still unable to address. Intuitively, it would be safe to assume that shorter words and sentences are easier to understand; however, lexical complexity is only one component of easability in reading comprehension and there are other factors such as word hyponymy and polysemy, argument overlap, and word meaningfulness that can all be factors complicating the comprehension of instructional and procedural texts for technicians with lower reading grade level (McNamara et al., 2011, 2012). Thus, it appears that the current simplification strategies, which mainly control and prescribe word choice and sentence length for MRO writing, may still not contribute significantly or adequately to increased comprehension of critical information by many global technicians.

Aviation technicians do not necessarily need intensive instructions on ASD-STE100 standards, but the final results of efforts to improve comprehension of manuals need to take into account all groups of users/readers. MRO manual writers and content creators, then, will have to be mindful of textual factors, including paragraph-level features of cohesion, to go beyond word-level needs, especially for global technicians. Werfelman (2007) notes that it is hard to determine how correct the content of texts is, but the ultimate goal of technical writing should be increasing the readability of texts while giving very clear and uncomplicated instructions. Of note here is the fact the English learned in schools in countries like China, Mexico, or Argentina relies heavily on academic writing that does not reflect or correspond to STE standards.

It is important for language trainers to provide technicians with consistent vocabulary instruction in context (Brown & Lee, 2015) and to reinforce their understanding of genre-specific writing styles of MRO manuals. Technicians' familiarity with how manuals link or connect ideas may better facilitate their technical comprehension and understanding of instructional language in these manuals. For corpus-related instruction, it may be helpful for instructors to provide trainees with concordances from the MRO manuals corpus that feature connectives used with approved ASD-STE100 dictionary words. For related material design, aviation textbooks and assessment materials for technicians will benefit from maintaining a focus on MRO manuals currently in use, especially on features that influence readability. Reading texts that match or are slightly above MRO manuals' reading difficulty level can be regularly utilized to develop enhanced reading strategies and abilities.

Chapter Co-Author

Genggeng Zhang is a doctoral student of Applied Linguistics and Asian Studies at Penn State University. She has a master's degree in Applied Linguistics from Georgia State University and a master's degree in Chinese Linguistics and Language Acquisition from the Chinese University of Hong Kong. She has been teaching Chinese and English since 2015. Her current research focuses on text readability, the linguistic features of text that affect comprehension, and their implications in language teaching.

6

Pilot-Controller Communication: A Multidimensional Analysis

6.1. Introduction

In addition to describing countable word or phrase-level features of written and spoken aviation discourse, presented in the previous chapter, it is also useful to examine these corpora more completely by analyzing internal patterns such as the distribution and use of co-occurring linguistic features of aviation texts from those annotated or "tagged" for part-of-speech (POS). This approach allows for register comparison—comparisons that will reveal how similar or different the discourse of aviation is in comparison with other domains, based on statistically defined linguistic parameters. To support the argument that the language of aviation is unique, with clear standards and conventions (i.e., as a *lingua franca*), it would be essential to establish its core features beyond characteristic words, phrases, or clauses and into the more complex building blocks of discourse. The goal here is to show broader patterns of language use, relative to other settings such as telephone interactions or office talk. A register comparison of this nature also has implications for assessment and for the development of teaching and training materials to address actual language use instead of following teacher or trainer intuition or standards proposed by companies or government (language-communication) monitoring agencies.

Many corpus-based studies often start with the extraction of frequencies of a small number of variables. Corpora can be grouped and defined to represent various registers and explored to produce word lists and collocations. With POS-tagging software, grammatical features can also be counted (e.g., counts for pronouns, passive verbs, and independent clauses), providing an additional set of linguistic data for interpretation and revealing the unique characteristics of a particular domain, aviation in this case. Questions such as "Is the use of 1st person pronouns in ego-centric sequences (e.g., *I think that it is needed...*) different across groups of pilots based on their first language background?" could be answered using this approach. Friginal and Hardy (2014b) argue that language use is often influenced by multiple contexts and variables and existing variations can be shown across combinations of discourse features distributed in a corpus. Sub-registers within a communicative context (e.g., oral communication) may influence the way speakers and writers use the many structures of a language available to them. How professionals write is clearly different from how they speak, and the purposes and audiences for such language production also affect the form of language they use.

In the Gladwell (2008) book mentioned in Chapter 4 and also from Chapter 1, intimidation as a potential factor influencing oral communication may have prevented the Colombian Avianca 052 co-pilot from more emphatically calling out the imminent emergency they were confronting by determining the form and structure of English he used in communicating with the JFK controller. A typical corpus will not capture segmental and suprasegmental features of pronunciation including, especially, tone, pitch, and volume, but other POS-tagged features may show personal references, respect markers, and passive verbs, even when they are not prescribed in the phraseology, to potentially reveal role and relationships and, especially, understanding of power and attitudes in the communication loop. Speakers are likely to adjust their language use when situations and people change. They may speak differently with peers than they do with employers or researchers who may be recording their responses. In order to further investigate these changes or accommodations in speech, potentially leading to countable linguistic variations, a single variable or feature (e.g., personal pronouns or the use of *please*) may not be enough to provide the more complete and useful picture of these kinds of variation in different contexts. Single counts for features can also be supplemented by other linguistic features that are used together or that co-occur with these words across registers and groups of speakers. The concept of **linguistic co-occurrence** explains that the linguistic composition of a sub-register such as face-to-face interaction may have higher frequencies of questions and responses, inserts, dysfluent markers (e.g., filled pauses—*uh, um*), and backchannels (e.g., *uh-huh*) used often by speakers in the corpus. Conversely, these features may not be common in written texts unless they come from character dialogues in fiction. Past tense verbs and nouns often go together whenever speakers engage in everyday conversations or talk about their previous experiences and recent, completed events. These same features could also appear together with very high frequency in written, 1st person narratives or soliloquies about past events. In order to capture and document these co-occurring features from corpora, a simple key word search will no longer be sufficient. A more advanced statistical framework is necessary to identify the composition of features that are frequently found together within a corpus.

6.1.1. The focus of this chapter

This chapter compares the functional features of "linguistic dimensions" from six spoken, professional, and primarily radio/telephone-based interactions, highlighting pilot-controller communication with other similar communicative domains: (1) customer service transactions from "call centers," (2) aviation interactions from pilots and ground controllers, (3) oral communication in the maritime industry, (4) telephone conversations between friends and family members, (5) spontaneous telephone exchanges between participants discussing topics identified by fixed prompts, and (6) face-to-face English conversation in the United States. These sub-corpora were taken from various sources including the Call Center corpus collected by Friginal (2008, 2009, 2013a), the Call Home corpus, and a sub-section of the Switchboard corpus (described below). The Call Home and Switchboard corpora were obtained from the American National Corpus (ANC) (see ANC's website at: http://www.americannationalcorpus.org/) and through the Corpus Linguistics Program at Northern Arizona University. Texts from face-to-face English conversation from the American English Conversation (AmE Conversation) corpus collected by Longman were used as additional comparison group. Linguistic comparisons across these registers followed a corpus-based, multidimensional approach developed by Biber (1988) using established dimensions of customer service talk from Friginal (2008, 2009, and 2013a).

6.2 Multi-feature, multidimensional analytical framework

Biber's (1988) multi-feature, multidimensional analytical (MDA) framework, from his groundbreaking book, *Variation across Speech and Writing*, has been applied in the study of a range of spoken and written registers and used in the interpretation of various linguistic phenomena. MDA data come from factor analysis (FA) which considers the sequential, partial, and observed correlations of a wide range of variables, producing groups of occurring factors or dimensions. According to Tabachnick and Fidell (2017), the purposes of FA are to identify and summarize patterns of correlations among variables, to reduce a large number of observed variables to a smaller number of factors or dimensions, and to provide an operational definition (i.e., a regression equation) for an underlying process by using these observed variables. The purposes of

FA support the overall focus of corpus-based MDA which aims to describe statistically correlating linguistic features and group them into interpretable sets of linguistic, functional dimensions. The patterning of linguistic features in a corpus creates linguistic dimensions which correspond to salient functional distinctions within a register, and allows cross-register comparison. MDAs of spoken registers have included topics such as gender and diachronic speech (Biber & Burges, 2001; Rey, 2001), stance and dialects of English (Precht, 2000), televised cross-cultural interaction (Connor-Linton, 1989; Scott, 1998), and job interviews (White, 1994). From 2008, several studies by Friginal on workplace spoken interactions have also been analyzed using MDA.

With computational tools such as grammatical POS-taggers and statistical tests that identify internal configurations of tagged data from corpora, it has become possible to establish the groups of co-occurring linguistic features and compare them across speaker demographics or sub-registers. Biber's MDA framework paved the way for many follow-up books, manuscripts, and journal publications, including a quantitative exploration of cross-linguistic comparisons of differing languages (Biber, 1995), as well as an edited volume of papers applying Biber's (1988) dimensions to more specialized corpora (Conrad & Biber, 2001). After thirty years since the publication of *Variation across Speech and Writing*, new sets of dimensions have been established by conducting new FAs across many specialized discourse domains in English, as well as many other languages. Cross-linguistically, the approach has been applied to analyze register variation in an equally extensive set of languages, including Somali (Biber, 1995; Biber & Hared, 1992), Korean (Kim & Biber, 1994), Spanish (Asención-Delaney & Collentine, 2011; Biber et al., 2006; Parodi, 2007), and Brazilian Portuguese (Berber Sardinha, Kauffmann, & Acuzo, 2014).

6.2.1. Friginal's (2008, 2009, and 2013a) dimensions of call center interactions

For the purposes of this chapter, linguistic dimensions from Friginal (2008, 2009, and 2013a) were used to compare the distribution of various vocabulary and grammar features of six groups of telephone-based registers, including one set of texts of face-to-face interactions. The composition of the tag-counted features for Friginal's FA was based primarily on prior studies, especially Biber (1988) of spoken and written English and White (1994) of professional job interviews. Additional discourse features of telephone-based service transactions (e.g., filled-pauses, politeness markers, length of turns and others) were included in the statistical dataset.

Comparing call centers with pilot-ATC talk: There are many similarities between pilot-controller interactions and those occurring in outsourced customer service call centers. Intercultural communication in customer service has become an everyday phenomenon in English-speaking countries like the United States, as callers (i.e., customers) come into daily direct contact with overseas-based call-takers ("agents"), typically from the Philippines and India. Before the advent of outsourcing, Americans had a different view of customer service facilitated on the telephone. Calling helpdesks or the customer service departments of many businesses mostly involved *local* call-takers and interactants typically shared the same "space and time" and contextual awareness of current issues inside and outside of the interactions themselves (Friedman, 2005). These forms of business transactions, therefore, have produced a relatively new register of cross-communication involving a range of variables not present in other globalized business or international and interpersonal communication settings. The same is true in aviation spoken discourse, with the growing number of international pilots regularly communicating with US-based air traffice controllers (ATCs) or international pilots in a country like China, communicating in English with Chinese ATCs. Interestingly, in both of these cases, there are prescribed norms for asking questions or responding, with varying levels and types of similarities/differences across contexts. Consequently, language factors and commonalities of space and attitudinal or cultural expectations are typically playing major roles in how interactions are conducted and completed, and how speakers maintain or deviate from prescribed norms (i.e., phraseology). Clearly, both of these domains have political, economic, and sociolinguistic implications impacting communication. In the case of call centers, the outsourcing of American jobs is an issue that has saturated the media and, in many cases, has also negatively influenced popular opinion, especially in the United States, affecting callers' attitudes and, consequently, their communication with "foreign" customer service agents and prompting calls from some sectors for policy changes and possible restrictions in business outsourcing practices (Friginal, 2013b). In aviation, an analogous dynamic may influence to some extent interactants' and also regulators' perceived understanding and rapport, and, in the case of the aviation domain, safety concerns, especially with the elevated role of English for non-native speakers and the characteristics of US ATCs as they perform their tasks using English native speaker norms and expectations.

Philippine and Indian Englishes in US call centers are in the forefront as American customers typically demand to be given the quality of service they expect or can ask to be transferred to an agent who will provide them the service and the "kind" or level of native language/English proficiency they prefer. Overseas agents' "performance" in language and explicit manifestations of pragmatic skills aligned with these native English speakers' expectations are, therefore, examined closely when defining "quality" during these outsourced call center interactions (Friginal, 2009, 2013b; Hayman, 2010). US-based ATCs are, of course, not customers, but in some situations, they do have the power and related attitudinal norms that make them expect and even demand very specific information, prescribe certain forms of language, and immediate action from international pilots, and—whether or not they're aware of it—they may be culturally and linguistically "pre-programmed" to expect certain characteristics in the communication they have with non-native English speakers. The native vs. non-native English speaker factors in some of these interactions provide evidence of these behavioral norms and expectations. A comparison of these two domains, together with other telephone-based and spoken corpora, may illustrate how similar or different these intercultural communication exchanges are and how interlocutors characteristically utilize statistically co-occurring features in common interactional exchanges. The implications of these sometimes varied cultural expectations and the resulting congruent or disparate expectations during communication are, obviously, profound within the context of the airline industry.

6.3. Methodology

Detailed theoretical and methodological explanation of the MDA approach can be found in Biber (1988), Friginal (2009), Friginal and Hardy (2014b), Conrad and Biber (2001), and Berber Sardinha and Veirano Pinto (2014, 2019). The following sections briefly summarize two interrelated MDA studies conducted by Friginal in 2008, 2013a, and 2015. His primary dataset came from a call center corpus, which he compared with other professional conversations conducted over the telephone. In this chapter, the same group of texts is also utilized, with the addition of two exploratory corpora of Aviation and Maritime interactions in English. Both Aviation and Maritime texts feature multicultural speakers, with native and non-native speakers of English coded in the texts.

6.3.1. Conducting FA of call centers (Friginal, 2008, 2013a, and 2015)

A total of 37 POS-tagged linguistic features shown in Table 6.1 comprised Friginal's MDA, after running various iterations and models from an original list that included up to 123 features. Discourse particles, 2nd person pronouns, average word length, total word count, length of turns, and type/token ratio loaded highly in the three factors. Friginal's (2008) FA reported that Kaiser-Meyer-Olkin Measure for Sampling Adequacy (KMO=.767, middling) and Bartlett's Test for Sphericity (Approx. Chi-Square=15121.685, df=757; p<.0001) were sufficient for exploratory FA with principal axis factoring. Results from a three-factor solution were deemed to be the most interpretable merging of features, with 37.97 cumulative percentage of Initial Eigenvalues (Total Variance Explained).

Table 6.1 Complete list of linguistic features used in Friginal (2008, 2013a, and 2015)

Linguistic Features	Description/Example
Type/Token	Number of words occurring in the first 400 words of texts
Word Length	Mean length of words in a text (in letters)
Word Count	Total number of words per agent/caller texts
Private Verbs	e.g., *anticipate, assume, believe, feel, think, show, imply*
That Deletion	e.g., *I think [Ø]she's away.*
Contractions	e.g., *can't, I'm, doesn't*
Present Tense Verbs	All present tense verbs identified by the tagging program
2nd Person Pronouns	*you, your, yours, yourself* (and contracted forms)
Verb *Do*	*do, does, did* (and contracted forms)
Demonstrative Pronouns	*that, those, this, these*
1st Person Pronouns	*I, me, my, mine, myself* (plural and all contracted forms)
Pronoun *It*	Instances of pronoun *it*
Verb *Be*	Forms of *Be* verb
Discourse Particles	e.g., *oh, well, anyway, anyhow, anyways*
Possibility Modals	*can, could, might, may*
Coordinating Conjunctions	*and, or, but*
WH-Clauses	Clauses with WH (*what, which, who*) head
Nouns	All nouns identified by the tagging program
Prepositions	All prepositions identified by the tagging program
Attributive Adjectives	e.g., *the small chair*
Past Tense Verbs	Past tense verbs identified by the tagging program
Perfect Aspect Verbs	Verbs in perfect aspect construction
Nominalizations	Words ending in *–tion, -ment, -ness*, or *–ity* (and plurals)
Adverb Time	Time Adverbials e.g., *nowadays, eventually*
Adverbs	Total Adverbs (not Time, Place, Downtoners, etc.)
Prediction Modals	*will, would, shall*

(Continued)

Linguistic Features	Description/Example
Verb *Have*	has, have, had (and contracted forms)
Average Length of Turns	Total number of words divided by number of turns
Filled-Pauses	uhm, uh, hm
Respect Markers	ma'am, Sir
Politeness Markers – Thanks	thank you, thanks, [I] appreciate [it]
Politeness Markers – Please	please
Discourse Markers – OK	ok (marker of information management)
Discourse Markers – *I mean*	*I mean* and *You know* (marker of participation)
Discourse Markers – Next/Then	next, then (temporal adverbs)
Discourse Markers – Because	because, 'coz, so (marker of cause and result)
Let's or *let us*	Instances of *let's* or *let us*

Table 6.2 Summary of the linguistic features of the three factors extracted from the Call Center corpus

Dimension	Features	
	Positive: Addressee-focused, polite, and elaborated information	
	2nd Person Pronouns	.683
	Please	.523
	Nouns	.515
	Possibility Modals	.445
	Nominalizations	.394
	Length of Turns	.376
	Thanks	.325
Dim 1:	Ma'am/Sir	.309
	⇕	
	Negative: Involved and simplified narrative	
	Pronoun *It*	–.687
	1st Person Pronouns	–.663
	Past Tense Verbs	–.609
	That Deletion	–.506
	Private Verbs	–.439
	Perfect Aspect Verbs	–.345
	I mean/You know	–.338
	Verb *Do*	–.321

Dimension	Features	
	Positive: Planned, procedural talk	
Dim 2:	Word Count	.821
	Length of Turns	.678
	Type/Token	.630
	2nd Person Pronouns	.515
	Next/Then	.417
	Word Length	.422
	Adverb Time	.409
	Prepositions	.383
	Please	.369
	Present Tense Verbs	.341
	Nominalizations	.321
	Because/So	.310
	Let's	.300
	⇕	
	Negative:	
	Discourse Particles	-.397
	Positive: Managed information flow	
Dim 3:	Discourse Particles	.947
	OK	.865
	Adverbs	.845
	Let's	.422
	⇕	
	Negative:	
	Length of Turns	-.349

After completing the final FA test, the composition of the three extracted factors (i.e., linguistic dimensions) of call center interactions from Friginal (2008) is presented in Table 6.2. Factor loadings and subsequent functional interpretations (i.e., factor names) of each dimension are also presented in this table and the following Results sections. The three functional dimensions of call center talk from Friginal's original factor analysis (of only texts from his Call Center corpus) are: (1) addressee-focused, polite, and elaborated information vs. involved and simplified narrative; (2) planned, procedural talk; and (3) managed information flow. The linguistic composition of these dimensions is enumerated in Table 6.2.

Table 6.3 Composition of corpora used in the present study

Corpora	Number of Text Files	Number of Words	Average Number of Words per Text File
(1) Call Center	500	553,765	1,108
(2) Aviation	860	361,200	420
(3) Maritime	211	51,695	345
(4) Call Home	120	345,237	2,876
(5) Switchboard	600	1,057,830	1,763
(6) American Conversation	200	1,166,105	5,828

6.3.2. Corpora analyzed in this chapter

Table 6.3 summarizes the composition of professional spoken corpora used for register comparison in this chapter. A brief description of these six corpora is provided below.

Call Center corpus

The corpus of call center transactions was collected by Friginal from 2006 to 2009 in the Philippines from three sponsoring call center companies all using web-based software for storing audio files of transactions for quality monitoring and documentation of transactions. The calls in the corpus ranged from five to twenty-five minutes in duration. The 500 audio files that comprise the Call Center corpus have an average call duration of 8 minutes and 45 seconds per transaction and have a combined length of over 120 hours of customer service interactions. Convenience sampling of audio files was done to ensure, among other considerations, a comparable number of files per task category (e.g., troubleshooting, telemarketing) or a balanced number of male and female call-takers and callers as much as possible. The audio files of customer calls were transcribed by trained Filipino transcriptionists following conventions used in the collection of the service encounter corpus of the TOEFL 2000 Spoken and Written Academic Language (T2K-SWAL) (Biber, 2006).

Call Home corpus

The Call Home corpus (or "Call Home English Corpus of Telephone Speech") consists of 120 unscripted and unplanned telephone conversations between native speakers of American English. All calls, mostly lasting up to thirty minutes in length, originated in the United States; however, 90 of the 120 calls were directed or placed to different locations outside of the United States. Most participants called family members or close friends following specific instructions and suggested topics developed by the research team during data collection. The Call Home corpus from the Linguistic Data Consortium (LDC) contains speech data files and minimal amount of documentation needed to describe the contents and format of speech files and the software packages needed to un-compress the speech data ("Call Home American English Speech," 2004).

Exploratory Aviation corpus

The exploratory Aviation corpus collected a range of text samples from various sources, including those provided by airlines operating in the Philippines, China, and Malaysia with service to US locations. Training and simulation texts from a leading South American airline are also included in the exploratory corpus, together with the Corpus of Pilot and ATC Communication or CORPAC from a corpus collection being conducted by Pacheco and Cavallet from The Pontifical Catholic University of Rio Grande do Sul, Brazil. CORPAC's primary data source is VASAvisation's YouTube channel (search for "vasaviation" from https://www.youtube.com), with publicly available audio files (most with accompanying transcripts) of authentic materials that feature a sampling of actual language used by pilots and ATCs in aviation in emergency situations. This corpus is relatively small, but growing, with over 50 events and close to 80,000 words. Recorded and transcribed interactions are coded manually to provide additional descriptive information as shown in the example below (Table 6.4).

Table 6.4 Sample CORPAC text with brief header information

Header Information	Cessna 150 Commuter (N150EC)
URL	https://www.youtube.com/watch?v=oi-LZkNb1yk
Title/Nature of the problem	Cessna C150 Commuter crashes; Location: Melbourne, FL
Date	December 19, 2016
Flight/Company/Aircraft	Cessna 150 Commuter/N150EC
Where (from/to)	Crash location: Eastern coast of Florida
Language, if known	Pilot—ESL/EFL; ATC—English first language
Phase of flight	Cruise

(Continued)

Header Information	Cessna 150 Commuter (N150EC)
Duration of transcript	2 minutes, 12 seconds
Summary	A Cessna 150 commuter (N150EC) aircraft on a flight over the eastern coast of Florida was forced to crash-land on a foliated rough pasture terrain. The airplane came to rest inverted, sustaining substantial damage, with two people on board, including the pilot suffering from unspecified injuries.

- ATC-Pilot Transcript -

MLB TWR: FIT one-three, Melbourne. Could I get you to fly to the abandoned for me and take a look at the abandoned to see if you see anything unusual there?
FIT 1–3: Anything unusual ... Yes, sir.
MLB TWR: Thank you.
FIT 1–3: Now, Tower, FIT one-three. Just- What kind of unusual activity am I looking for?
MLB TWR: They called from the Brevard County Fire and Rescue and I want to see if there's anything on the ground there I need to worry about.
FIT 1–3: Roger that. We're keeping an eye out. I'll tell you in a bit shortly.
FIT 1–3: Tower, nothing unusual that I can see. But I am just seeing a vehicle actually driving on the premises.
MLB TWR: OK. But you don't see any aircraft in distress or anything like that going on over there?
FIT 1–3: I do not see anything in distress, no fire, no anything. Just, like I said, one vehicle just driving on the paved area.
MLB TWR: OK. Thank you, sir. You can fly straight into runway five; report passing the interstate.
FIT 1–3: Alright. Straight runway five, report interstate. FIT one-three.
N789BG: Melbourne Tower, November-seven-eight-nine-Bravo-Golf. ILS niner right, requesting ... [unintelligible].
MLB TWR: November-seven-eighty-nine-Bravo-Golf, Melbourne Tower. Report Passing the interstate highway.
N2380A: Tower, Cessna two-three-eight-zero-Alfa with you and we have Foxtrot.
MLB TWR: Alright, I can only talk to one of you guys at a time. Somebody calls, you gotta let the other guy answer. Nine-Bravo-Golf, report passing the interstate.
N789BG: Report passing the interstate. Seven-eight-nine-Bravo-Golf.
MLB TWR: OK. Who called in while he was answering?
N2380A: Cessna two-three-eight-zero-Alfa. We have Foxtrot, and we're with you.
MLB TWR: Two-three-eight-zero-Alfa, roger. And what are your intentions.
N2380A: *Yeah*, sir. We'd like niner right and *uh*- [unintelligible] the southeast.
MLB TWR: Cessna two-three-eight-zero, fly straight in runway nine right. Report passing the interstate.

- End of transcript -

Maritime corpus

The texts from the Maritime corpus were provided by four European shipping companies employing a large number of merchant marines primarily from Asian countries. A majority of interactions came from carriers that travel from European and Asian port locations to various US destinations. The corpus has 211 (51,695 total words), coded to include speaker roles and job functions. Several of the texts include communications between ship personnel and port authority staff, which may resemble pilot-ATC interactions.

Switchboard corpus

The Switchboard corpus is composed of spontaneous conversations of "telephone bandwidth speech" between US-based speakers. The corpus was collected by Texas Instruments and funded by the Defense Advanced Research Projects Agency (DARPA). A complete set of Switchboard CD-ROMs available from the Linguistic Data Consortium includes about 2,430 conversations averaging six minutes in length (with over 240 hours of recorded speech), and about 3 million words of text, spoken by over 500 speakers of both sexes from every major dialect of American English ("Switchboard: A Users' Manual," 2004). A total of 600 files with over approximately 1 million words comprise the Switchboard sub-corpus used in this chapter. Interaction with the switchboard system was conducted via touchtones and recorded instructions given to the participants. The topics for conversation (e.g., *"What do you think about dress codes at work?"* or *"How do you feel about sending an elderly family member into a nursing home?"*) were randomly identified by the system. The two speakers, once connected, were allowed by the system to "warm-up" before recording began. The speakers did not know each other personally and had no previous information about each other's personal backgrounds before the warm-up conversation.

American English (AmE) Conversation corpus

The American English Conversation corpus used in this chapter was obtained from the Longman Grammar Corpus of Spoken American English. The Longman Grammar corpus has over approximately 4 million words and was designed to be a representative corpus of American conversation covering a wide range of speech types (e.g., casual conversation, service encounters, task-related interaction), locations or settings (e.g., home, classroom), geographic regions in the United States, and speaker characteristics (e.g., age, gender, occupation). Only text files of face-to-face conversations from this corpus were used in the present study. The American Conversation sub-corpus has a total of 200 texts with approximately 1.1 million words.

6.3.3. Computing dimension scores

After compiling the corpora, all individual texts were tagged for linguistic features, especially part-of-speech, using the Biber grammatical tagger. The frequency counts of a total of 137 POS features were normalized per 1,000 words across texts. The current version of the Biber tagger incorporates the corpus-based research carried out for the *Longman Grammar of Spoken and Written English* (Biber et al., 1999). This tagger identifies a range of grammatical features, including word classes (e.g., nouns, modal verbs, prepositions, verbs), syntactic constructions (e.g., WH relative clauses, conditional adverbial clauses, *that*-complement clauses controlled by nouns), semantic classes (e.g., activity verbs, private verbs, likelihood adverbs), and lexical-grammatical classes (e.g., *that*-complement clauses controlled by mental verbs, *to*-complement clauses controlled by possibility adjectives).

The frequencies of co-occurring linguistic features from the three dimensions (Table 6.2) across all corpora were standardized (using z-scores), allowing greatly differing distributions to be more comparable with each other and offering scores that reflected a feature's range of variation. Each dimension comprised linguistic features that significantly co-occurred with one another and contained both positive and negative loadings. Standardization of frequencies allowed for these complementary patterns of polarity. In other words, when a text contains frequent instances of one group of co-occurring linguistic features (positive or negative), the features from the opposite group are likely to be very limited (Biber, 1988). Using the composition of Friginal's dimensions, the standardized frequency data (z-scores) from all six corpora in this chapter were then added to obtain dimension scores per individual text. Once scores in all three dimensions had been calculated for each text, mean scores per corpus were obtained by averaging the texts' dimension scores.

6.4. Results

For each of the three dimensions, six average scores comprising the corpora compared in this chapter are shown along comparison figures below. These figures describe cross-register linguistic distributions and relationships per dimension. Text samples with high or low dimension scores focusing especially on pilot-ATC interactions are provided in the following sections to better reveal and understand the functional characteristics and the significance of these distributions.

6.4.1. Dimension 1: Addressee-focused, polite, and elaborated information vs. involved and simplified narrative

Eighteen linguistic features comprise this dimension with nine features on each of the positive and negative sides. Positive features include politeness and respect markers (e.g., *thanks, please, ma'am* and *sir*), markers of elaboration and information density (e.g., long words and turns, nominalizations, and more nouns), and 2nd person pronouns (e.g., *you, your*) which indicate "other-directed" focus of talk. Possibility modals (*can, could, may, might*) also loaded positively on this factor. The features on the negative side of this factor, especially pronoun *it*, 1st person pronouns, *that* deletion, private verbs, WH-clauses, and verb *do*, resemble the grouping in the dimension "Involved Production" identified by Biber (1988) and White (1994). These features are typical of spoken texts and generally contrast with written, informational, and planned discourse. Also on the negative side of the factor are past tense verbs, perfect aspect verbs, and the use of discourse markers *I mean* and *You know*. These elements point to an accounting of personal experience or narrative that tries to explain the occurrence of a particular situation or event. Schiffrin (1987) considers *I mean* and *You know* as markers of information and participation; *I mean* marks speaker orientation toward the meaning of one's own talk while *You know* marks interactive transitions.

These co-occurring sets of features represent the contrast between the dominant objectives of speakers' utterances. Speakers in telephone exchanges who use more positive features are likely intending to give details, explanations, or solutions (especially in the case of customer service call-takers). In the process, these interactants use more nouns, nominalizations, and longer utterances or turns to deliver the information. The information density in these turns is high because of higher average word lengths in the texts. Participants' turns are elaborated with detailed explanations, likelihood, or risks through the use of a significantly high frequency of possibility modals. The high frequency of 2nd person pronouns indicates that the transfer of information is highly addressee-focused.

Conversely, the grouping of features on the negative side of the dimension illustrates personal narrative and experiences, and simplified information. The combination of past tense verbs, private verbs, pronoun *it*, and discourse markers *I mean* and *You know* demonstrates the typical goal of utterances which is to provide a personal account of how a situation or an event happened. Involved production features (e.g., 1st person pronouns, WH-clauses, verb *do*, and *that* deletion) and *I mean, You know* serve a communicative purpose to establish personal orientation (White, 1994) and purposely ask for a response. Most utterances on the negative side of the dimension have fewer word counts and are significantly shorter in length. To summarize, the combination of positive and negative features of Dimension 1 differentiates between addressee-focused, polite, and elaborated information and involved and simplified narrative portraying how informational content is produced in the discourse. Figure 6.1 shows the range of variation across the six corpora.

Call Center (4.77), Maritime (2.47), and Aviation (0.22) texts have scores on the positive side of the dimension while the more informal telephone and face-to-face conversation corpora have average dimension scores on the negative: Call Home (−0.98), Switchboard (−1.78), and AmE Conversation (−4.56). Telephone service encounters commonly allocate for courteous language and

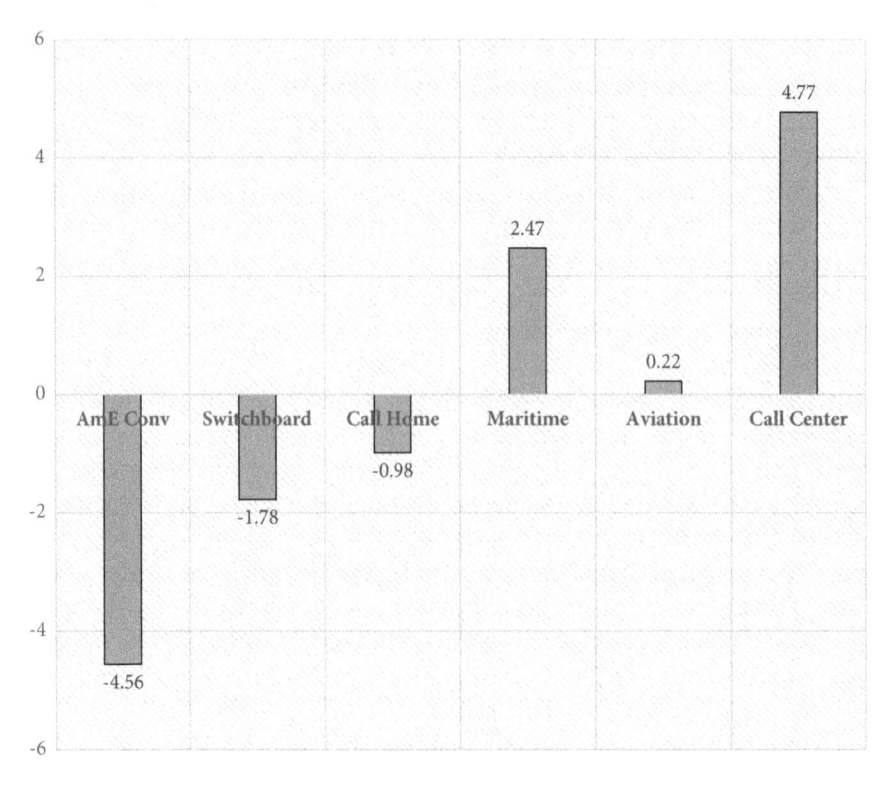

FIGURE 6.1 *Comparison of dimension scores for Dimension 1: Addressee-focused, polite, and elaborated information vs. involved and simplified narrative ANOVA: Registers, F=5.212; p<.001.*

the recognition of roles; call-takers, especially, are expected to show respect and courtesy in assisting their customers (D'Ausilio, 1998). In this dimension, the frequency of politeness and respect markers in the Call Center corpus is significantly higher than the other five comparison corpora. In fact, these features were seldom used in face-to-face texts. Both Biber (1988) and White (1994) characterize spoken discourse as highly involved and interactive from increased use of pronouns, private verbs, and discourse markers. Linguistic features that show spoken narratives (e.g., past tense verbs, pronouns, *that*-deletion, etc.) are also very common in these interactions especially in face-to-face conversations and also in Call Home (e.g., narratives and accounts of experiences or events).

Aviation texts average just slightly above zero in Dim 1, distinguishing them significantly from Maritime and Call Center discourse. In this dimension, plain language and specific phraseology expected or required in pilot-ATC communications contribute to the structure of the interactions. Polite markers (especially *please, thanks/thank you, appreciate, sorry, apologize*, etc.) are all very common in Call Centers, but not in Aviation. Maritime discourse has a significantly higher frequency of *sir* (as polite or respect marker) across the board, even more than Call Centers (with more *ma'am*). In general, following distributions consistent with the positive side of Dim 1, Maritime discourse is more polite than Aviation and more cognizant of roles and relationships, especially in recognizing ranks or positions (e.g., captain, officer, border patrol).

The three text excerpts below highlight the use of past tense verbs and personal pronouns (i.e., narrative features) in face-to-face conversations compared to polite, elaborated, and informational utterances from a call-taker in customer service interactions, and how these characteristics are not present in typical pilot-ATC exchanges (Text Sample 6.3). The Call Center text (Text Sample 6.1: task—purchase Mobile Phone Minutes, Dim Score = 6.113) shows detailed explanation and additional information given to the caller. Technical information, business-related items, and politeness markers are all used by the call-taker in this excerpt. The call-taker engages the caller by using conventional customer service responses (e.g., "*I apologize for the inconvenience..,*" "*thank you for calling ... how may I assist you*," or "*Let me just verify the charges..*"). In Text Sample 6.2 (setting: office/break time talk, Dim Score = –7.231), the two speakers discussed two overlapping sets of past events (bachelor party and previous work experience in North Carolina).

Text Sample 6.1. Purchase mobile phone minutes (Dim Score = 5.713) (name replaced by pseudonym)

Agent: **Thank you for calling** [company] Activation Center, this is [agent name], **how may I assist you?**

Caller:	Oh, yes, ah, I started to activate this phone before and I messed up on it, and I said I'll call back. Ah, my name is [caller name]
Agent:	**Ok, sir, thank you, please** give [interruption]
Caller:	I have a number that they gave me and I get it, uh, completely, uh complete activation.
Agent:	Uh, so we, **I do apologize for the inconvenience sir, can I have, please,** the uh cell phone number?
Caller:	Um, ok I have the new number that they gave me, it's [number].
Agent:	**Ok, sir,** that is [number]?
Caller:	That's it.
Agent:	**Ok, thank you for that information sir. Let me just check that number for you and let me just verify if there are charges.** This is the um, uh, the new number that they gave you **sir, is that correct?**
Caller:	Yes. I had an old number I was trying to get my minutes from, if you need that.
Agent:	**Ok sir.**

Text Sample 6.2. Talk during lunch time (Dim Score = −6.231)

Speaker 1:	It's like Greg um, we **had** the lovely bachelor party at our house for a friend and I **was** like fumigating my house when it **was** over.
Speaker 2:	Cigars?
Speaker 1:	Oh that **wasn't** it. My towels **smelt** bad. But it **wasn't** the cigars, it **was**, I could **have handled** <unclear>, it **was** bad stuff. And you know that **wouldn't have even bothered.** I mean I can handle that.
Speaker 2:	Excuse me?
Speaker 1:	Oh, oh that **was**, it **was** really, really and the thing **was** … well not … uh, let me look at your beer menu if that's here.
Speaker 2:	We'll, we'll snooze through the movie this afternoon but hey that's okay.
Speaker 1:	We'll find out so Greg **told** me … **told** me in the house right? He **told** me, this **happened** a year ago, he **told** me … I **had** uh, when I **checked** into my first duty station in the service I **was** in North Carolina and uh, fortunately probably **was** one of the finest working experiences I've **had** in my whole working life. They **were** real serious

	about their work, **took** it very seriously but they also **didn't** take themselves too seriously. And it **was**, senior NCO's down to the individuals **had** fun and yeah.
Speaker 2:	I'm watching to see if it does. It's got the record level on.
Speaker 1:	No I don't think so. Back up here. Everything's good?

Text Sample 6.3. Aviation (Dim Score = 0.012)

Description:	An American Airlines Boeing B737-823 (N807NN), flight AAL2405 from Miami KMIA to Antigua TAPA on departure when ATC reported that a passenger called 9–1-1. The aircraft returned and taxied to the gate, followed by emergency/police vehicles.
MIA DEP:	American twenty-four-zero-five, climb and maintain one-six, sixteen thousand. Proceed direct CRABI.
AAL2405:	One-six thousand, direct to CRABI. American twenty-four-O'-five.
MIA DEP:	Five, *er*-They appear to have overdosed. I have the passenger's number- I'm sorry- Name. Advise what you need.
AAL2405:	Say again for American twenty-four-O'-five.
MIA DEP:	American twenty-four-zero-five, a passenger from your flight called 9–1-1, said they overdosed and are on your flight. Now, the name is ***. Climb to one-two, twelve thousand.
AAL2405:	Twelve thousand. American *uh*-Twenty-four-O'-five. (…) *Uh*-Right now, can you spell the name? And do you know the seat number?
MIA DEP:	Delta-Alfa-[unintelligible]-Alfa-Whiskey-[unintelligible] Sierra-Echo.
AAL2405:	Right one-eight-zero. American twenty-four-O'-five.
MIA DEP:	And American twenty-four-zero-five, they think she may be kidnapped. The- She has - She's on the phone whispering and somebody's trying to take the phone away from her. (…) I don't have the seat number.
AAL2405:	OK. Copy that.
MIA DEP:	Four-zero-five, descend and maintain one-zero, ten thousand.
AAL2405:	Down to one-zero, ten thousand. American twenty-four-O'-five. (Second pilot, background: "Ten- ten thousand").
MIA DEP:	American twenty-four-zero-five, turn heading two-seven-zero; descend and maintain one-zero, ten thousand. Expect the ILS runway nine approach.

AAL2405: Two-seven-zero for American two-four-zero-five, down to ten.
MIA DEP: And American twenty-four-zero-five, were you able to find any information on your end?
AAL2405: Negative for American twenty-four-O'-five. We have no indication on our flight.
MIA DEP: Do you have any passenger by that name?
AAL2405: We cannot find anyone by that name.

In these three distinct interactions, overlapping linguistic markers of turn-taking and question-answer sequences are present, with both Call Center and Aviation exchanges conducted by means of radiotelephony, addressing distance and differences in speaker locations. Question-answer sequences are clearly marked by specific and very particular lexicon in Aviation and also Maritime interactions (e.g., *negative, say again, copy that*), not present in typical business and informal spoken corpora. There are only very few traces of narrativity in Aviation compared to typical spoken interactions, making Aviation texts mirror the structure of written documents in Dim 1 (close to zero in dimension score). This phenomenon is observed in most procedural turns, especially in simple instructional utterances and responses. Pilots are required to repeat or confirm understanding, and ATCs follow consistent sequencing of required call parts or sections, for example, (1) American twenty-four-zero-five; (2) turn heading two-seven-zero; (3) descend and maintain one-zero, ten thousand to provide specific instructions.

The variation in the use of features in Dim 1 appears to be influenced by the nature of the task or issue of concern confronting the caller in Call Centers and specific topics of discussion in informal spoken discourses. In Call Centers, callers with complicated issues and those who are reporting dissatisfaction with service tend to have increased frequencies of nouns, are more addressee-focused, and, as expected, have longer utterances or turns (e.g., the use of more nominalizations, nouns, and 2nd person pronouns in Text Sample 6.1). On the other hand, callers with less-problematic issues rely on the response from the agents once the context of the call has been established. Callers who need specific information such as pricing, product number, or dates have fewer nouns and relatively shorter utterances. Turns are limited to short questions and clarification sequences. For Aviation, a majority of the interactions are routinized and repetitive, making the use of negative and positive features fairly even. However, as discussed later in this chapter, some observable changes happen when pilots and ATCs are confronted by emergencies or when there are attitudinal factors involved in the exchanges.

6.4.2. Dimension 2: Planned, procedural talk

The items loading on the positive side of Dim 2 include lexical specificity and information density features (type/token ratio, average word length), temporal adverbs (*next/then*) and specific time adverbials (e.g., *eventually, immediately*), complex and abstract information features (word count, length of turns, and nominalization), 2nd person pronouns, prepositions, cause and result discourse markers (*because/so*), politeness marker *please*, present tense verbs, and *let's* (including *let us*). Only discourse particles (e.g., *oh, well, anyway*) loaded on the negative side. The positive side of the dimension, thus, signifies a one-way (addressee-focused) transfer of a large amount of abstract and technical information. In this case, the information is "real-time," procedural or process-based due to the presence of temporal adverbs combining with the imperative *let's*, prepositions (e.g., *in, on, below, above*), and, especially, present tense verbs. The frequent occurrence of present tense verbs in the texts with higher positive loadings illustrates the use of directives/imperatives in utterances (e.g., *"..then hit save"*; *"..now, remove the tracking tape ... "*). It appears that this form of instructional language, especially common in call center talk, is expressed through a series of directions marked by 2nd person pronouns (especially *you* and *your*), succession between steps (*next/then*) and progression through the discourse (*now*). Discourse particles, used very sparingly in this dimension, perhaps indicate that the utterances are somewhat prepared or organized, and produced with limited hesitations or tentativeness.

As shown in Figure 6.2, Dim 2 differentiates Aviation, Call Center, and Maritime interactions from the three other comparison spoken corpora. Face-to-face interactions, Switchboard discussions of topics, and telephone interactions between family members all have negative aggregate scores. These three informal telephone and face-to-face conversation corpora have a higher frequency of discourse markers which are in complementary distribution with temporal adverbs, cause and result discourse markers, and especially imperative *let's*.

Friginal (2009) suggests that the merging of linguistics features on the positive side of Dim 2 indicating lexical specificity, complexity, and abstraction of information helps to differentiate call center discourse (and in this chapter, transactional talk from Maritime and Aviation texts) from general conversation and other sub-registers of telephone-mediated talk. In typical customer service calls, for example, longer words (based on average word lengths) and technical vocabulary are often used in extended turns during the interaction. Information-packaging in call center discourses is, therefore, somewhat more similar to written, planned texts because of the presence of features that are not commonly produced online, including nominalizations and higher type/

token ratio. Biber (1988) states that these features are more common in academic written texts and less frequently observed in spoken texts because of the influence of production circumstances. In typical, online conversations, general topic shifts allow for the occurrence of more common words and phrases and limited complex or abstract vocabulary.

The Aviation corpus has a collective Dim 2 score of 3.67, 6.76 from the Call Center corpus, and 3.45 from the Maritime corpus. Clearly, specific procedures and instructions are not common in face-to-face interactions unless they involve the performance of tasks. In Switchboard, there are instances of short, procedural discourse especially in the beginning of the discussions when participants talk about the instructions following the automated prompts during the recording of their conversations. However, these instructions echoed by the speakers are also limited and not extensively repeated in the exchanges. Texts from Call Home and AmE Conversations have significantly higher frequencies of discourse particles such as *oh, well,* and *anyway/s* that are all very rare in aviation exchanges.

Procedural and instructional utterances tend to double in Aviation texts due to the common convention of communication confirmation or readback, as pilots acknowledge and repeat what ATCs have spoken to them as direct instructions. Many similar instances are also utilized in Maritime communications. Communication confirmation is critical during very specific parts of Aviation talk in take-offs and landings as interlocutors complete the process and make sure that all safety procedures are strictly followed. In all, the language of ATCs across registers has the highest average frequency of the co-occurring features

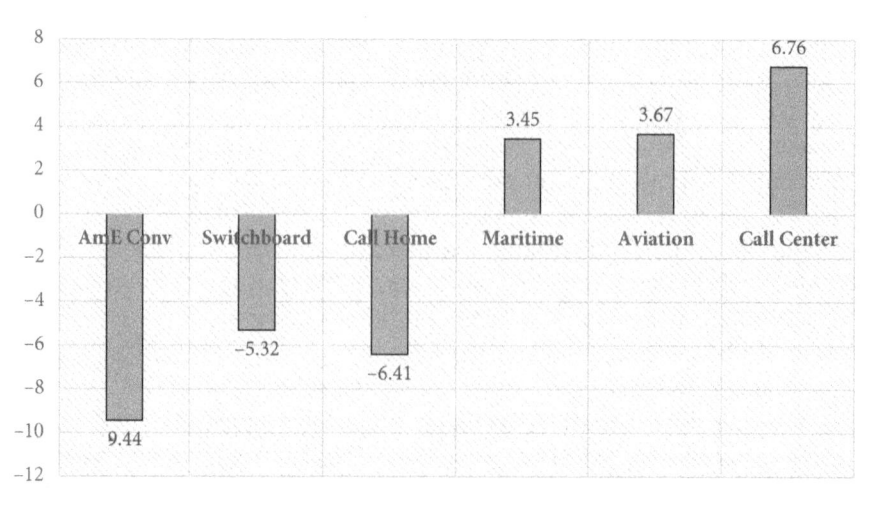

FIGURE 6.2 *Comparison of dimension scores for Dimension 2: Planned, procedural talk ANOVA: Registers, F = 30.134; p<.0001.*

of Dim 2 (positive side), even more than call center agents. In call centers, these call-takers use similar high-frequency features on the positive side Dim 2 and, predictably, engage in directive, procedural talk with their callers, but they also utilize other structures in their utterances as there are more question-answer sequences requiring extended explanations that may no longer be procedural or instructional. Call-takers' speech in this dimension is produced online but covers a wide range of topics and makes use of a variety of specialized terms or jargons that comprise their set scripts. Similar to ATCs, call-takers' utterances while giving directions and steps are planned, many of them written, because they have clear and, in many cases, anticipated expectations about the variety of caller questions they respond to. The moves in assisting callers are well-defined, and procedures are commonly established during many on-the-job training programs. For example, memorized procedural scripts (e.g., " ... *thank you for your call, first I will ask you for your account number...*") are often part of call-takers opening sequences from prescribed protocols. These memorized turns are also very common in Aviation and Maritime texts. One major difference in instructional talk between Call Center texts and Maritime/Aviation texts is the use of the imperative *let's* (or *let us*), highly frequent in customer service but almost non-existent in Aviation and Maritime discourses.

The collected text samples below (Text Samples 6.4) illustrate ATC instructional language, with some communication confirmation provided by pilots.

Text Samples 6.4. Procedural or instructional language by ATCs in the Aviation corpus

MIA APP:	American twenty-four-O'-five, **descend** and **maintain** three thousand.
AAL2405:	Down to three thousand now. American twenty-four-O'-five.
MIA APP:	American twenty-four-O'-five, **reduce speed** to one-nine-zero; **fly** heading three-zero-zero.
AAL2405:	One-nine-zero *uh*-Three-zero-zero for American twenty-four-O'-five.
MIA APP:	Yes, speed one-ninety, heading three-zero-zero. (...) American twenty-four-O'-five, descend and maintain two thousand. And did you get your gate yet?
AAL2405:	That's affirmative, we're at Delta four. And down to two-thousand. American twenty-four-O'-five.
MIA APP:	American twenty-four-O'-five, when you have time, fuel remaining and number of people on board.
AAL2405:	We got twenty-seven thousand, eight hundred pounds of fuel and we have ahundred forty-four souls on board. (...) For American twenty-four-O'-five.

***

Description: An Embraer Phenom 300, flight EJA386 from Santa Ana KSNA to Spokane KGEG suddenly lost complete control of the aircraft when cruising at FL400.
A few moments later, pilots regained control and were diverted to Reno KRNO for safe landing.

OPS 8:	Reno, I'd like to go off Alpha across one-six right at Delta and hold for the alert.
RNO GND:	Ops-eight **proceed as requested. Cross runway one-six right**.
OPS 8:	Ops-eight cleared to cross one-six right at Delta. We'll hold short of one-six left. Once the alert aircraft crosses, runway will be my control.
RNO GND:	Ops-eight, roger.
OPS 8:	And will it be next to land?
RNO GND:	Ops-eight *uh*-, **negative. We'll keep you advised**. We don't see it - even have him on radar just quite yet, but we are informed by NorCal that they were about ten minutes out five minutes ago.
OPS 8:	OK, thanks. And, **if you could, please update** on this frequency as to him regaining control. **Thanks**.
RNO GND:	Ops-eight, WILCO. When we last heard, he had regained control of the aircraft but there was a moment in time when he had lost complete control of the aircraft.
OPS 8:	Ops-right, copy that!
FIRE 4:	Reno Ground, Fire-four.
RNO GND:	Fire-four, Reno Ground.
RNO GND:	Emergency inbound is two-five, twenty-five miles north of the airport. **We'll keep you advised when he's the next arrival.**
FIRE 2:	And Reno Ground, Fire-two. Be advised: all fire equipment are now staged for one-six left.
RNO GND:	Fire-two, thank you.
NorCal APP:	Execjet three-eighty-six, **cleared visual approach** runway one-six left.
EJA386:	**Cleared the visual one-six left.** three-eighty-six.
NorCal APP:	Execjet three-eighty-six, **contact Reno Tower** one-one-eight point seven.
EJA386:	Eighteen seven. Thanks for the help. Execjet three-eighty-six.

RNO GND:	Inbound has two persons on board and three thousand, one hundred and seventy pounds of fuel.
RNO TWR:	... Tower. Wild calm. Runway one-six left, cleared to land.
EJA386:	One-six left. Cleared to land. Execjet three-eighty-six.
RNO GND:	All emergency equipment be advised: next aircraft to arrive runway one-six left is the alert aircraft.
FIRE 2:	Fire-two, copied.

Text Sample 6.5 (Dim Score = 8.333) shows an excerpt of planned, procedural interaction in a troubleshooting transaction from the Call Center corpus. This excerpt shows a range of new, technical words (e.g., *T1, DSL, Voice Over IP, broadband*) and nominalizations (e.g., *documentation, possibility, connection*) that are not necessarily repeated again in the text. The use of these words increases type/token ratio, average word count, and average length of turns in procedural interactions. In Call Center texts, temporal adverbs are very common (highlighted in bold below) but these features are actually discouraged in Aviation or Maritime phraseology, even when they seem to be needed, consequently, for the most part, prioritizing the use of specific (imperative) verbs such as *proceed, descend, reduce* [speed]. However, in the excerpt from Text Samples 6.4 (below), the ATC's request or instruction to the pilot to provide "fuel remaining" and information on the "number of people on board" did not make use of an imperative verb (e.g., "... give me fuel remaining and number of people on board").

MIA APP—	American twenty-four-O'-five, when you have time, **fuel remaining** and number of people on board.

Text Sample 6.5. Troubleshooting interaction from the Call Center corpus (Dim Score = 8.333) (caller's name is a pseudonym)

Call-taker:	**Then go ahead** and please type in "Yes" and **then hit 9**
Caller:	Ok, and then enter again?
Call-taker:	Yes, uh-huh?
Caller:	Well it just says dialing
Call-taker:	Uh-huh, by the way Sarah **just give me an update** whenever the message on the screen changes so that I could uh put down **documentation** here
Caller:	Ok [long pause] it says "connect phone cord and press," then it says "done press enter"
Call-taker:	Hmm, it, it actually means Sarah that uhm the only reasons that the **postage machine** would say connect the "connect phone cord message" is because it's not **detecting** a

	dial tone because it's connect, it's hooked up to a wrong type of **phone line** or the **phone cord** itself is defective. **Now we need a connection**, uhm since this is a brand **new postage machine** uh there's a big possibility that the **phone line** that it's hooked up to is not correct, so uhm Sarah is it ok if I get the **phone number** where you have the **postage machine** hooked up to so that I could check if uhm if it's dialing out or not?
Caller:	Yeah it's the office number
Call-taker:	Are you on the same line as the **postage machine**?
Caller:	Uhm well it's actually connected to a connector, well there's three of them
Call-taker:	Oh you mean a splitter?
Caller:	Yeah
Call-taker:	**Now** that's actually the reason why it's not uh going out properly. As I said earlier uhm Sarah this **postage machine** needs a **dedicated analog line**, so when we say it's a **dedicated** line it should not be sharing the line with any other **equipment**, it should not have a **rollover system**, uhm if the number has **extensions** uh we should be sure that those uh **extensions** doesn't have any **equipment** hooked up to it, and uh when we also say **analog** we **have to make sure** that it doesn't have **T1, DSL, Voice Over IP**, or even **broadband** on it. Now the best example for a **dedicated analog** line would be your **fax line**, so if we could just [interruption]

In contrast to transactional texts, Text Sample 6.6 from Switchboard (Dim Score = −4.212) features spontaneous discussion about the weather with short turns and some highlighted use of several discourse markers, especially *well* and *oh*, again, not common in Aviation and Maritime exchanges.

Text Sample 6.6. Switchboard—"weather" (Dim Score = −4.212)

Speaker 1:	**yeah** we set a record yesterday and uh very very windy but then today the wind has dropped off and also the temperature so
Speaker 2:	**oh** very cool uh I think right now it's like **oh** sixty-nine
Speaker 1:	hm
Speaker 2:	and that's cool for **anyways**
Speaker 1:	or if it it feels cool compared to yesterday **though** but very pleasant no rain in the last month I don't think ground's very dry and

Speaker 2:	our yard work everything everything is in bloom **right** so our yard work's pretty tough uh ground being dry but
Speaker 1:	I guess **well** it also uh brings about allergies we're having a lot of allergies down here right now
Speaker 2:	um-hum
Speaker 1:	everything blooming and and the weather and uh think a lot of people have contracted uh spring fever
Speaker 2:	too so had a lot of people out at work **well you know** for fishing and and uh golf reasons and things like that
Speaker 1:	hm
Speaker 2:	the blue flu yeah
Speaker 1:	**yeah** the blue flu or the white collar flu depending on where you work I guess
Speaker 2:	**oh** we have had uh as I've said we've had variable weather uh
Speaker 1:	hm
Speaker 2:	it has been
Speaker 1:	untypically wet for this time of year
Speaker 2:	hm
Speaker 1:	and also we have a lot of
Speaker 2:	**oh** green **you know** the grass has been growing and **well**
Speaker 1:	if you look outside you
Speaker 2:	would like to go out and mow your lawn if you could go out and
Speaker 1:	spring and **well** I guess we're still in winter and uh

6.4.3. Dimension 3: Managed information flow

The linguistic features on the positive side of Dim 3 are discourse particles (e.g., *oh, well, anyway*), the discourse marker *ok*, occurrences of *let's* (and *let us*), and adverbs (any adverb form occurring in the dictionary, or any form that is longer than five letters and ends in *–ly*). The adverbs comprising this list do not include time and place adverbials and those counted as amplifiers or downtoners. The positive features in this factor are very common in most spoken registers, especially conversations.

Discourse particles are regarded as necessary for conversational coherence (Schiffrin, 1994) and in monitoring the flow of information in talk (Biber, 1988; Chafe, 1985; Friginal, 2009). *Ok* is also regularly used in conversation and purposeful interactions like service encounters, and it serves as either a marker of information management (Schiffrin, 1987) or an apparent backchannel (Tottie, 1991). The use of the imperative *let's* is characteristic of interactions that especially focus on the performance of tasks (Friginal, 2009). The combination of discourse particles and backchannels could be interpreted as a conversational device to maintain and monitor the overall flow of transactions. More of these features emerge because the interactions are conducted over the telephone with clearly defined turns and adjacency pairs. It is possible that backchanneling through *ok* and the use of discourse particles that initiate turns are preferred by participants in telephone interactions to avoid dead air or very long pauses.

Thus, the grouping of linguistic features in Dim 3 signifies speakers' attempt at managing the flow of information. In Call Center talk, for example, this dimension separates callers and call-takers in their use of discourse particles, *ok*, and adverbials intended to facilitate and monitor the transaction. Figure 6.3 shows that informal spoken interactions and Call Center texts all have positive average dimension scores in Dim 3, with Call Center having the highest average frequencies of discourse markers and *ok* (but both of these features are also commonly used in Switchboard and Call Home interactions). The use of *let's* contributes to the difference in the factor scores of the Call Center corpus compared to the other telephone-based corpora. There is a high frequency of *let's* and *let us* in the turns of call-takers in call centers

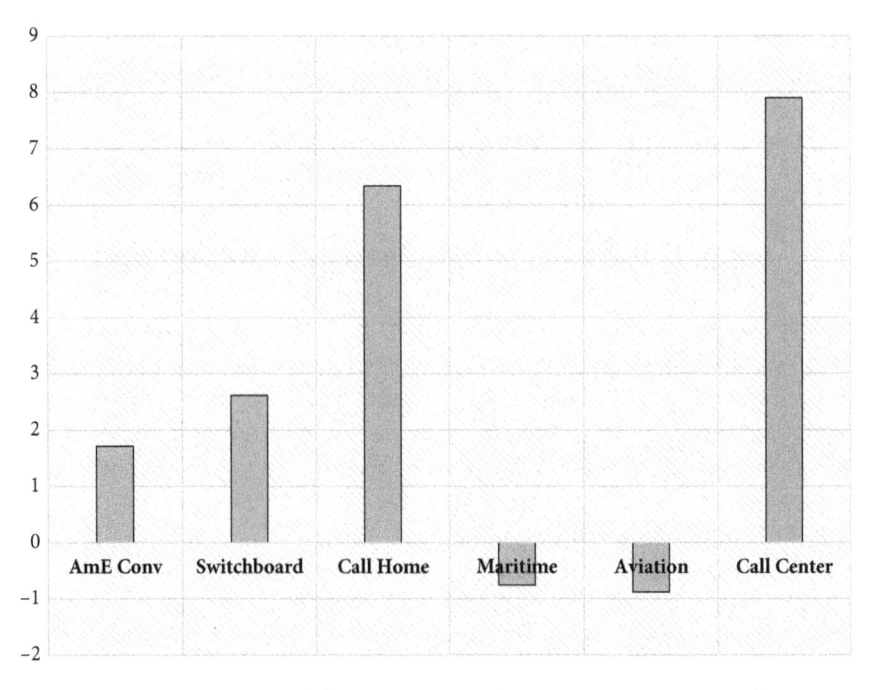

FIGURE 6.3 *Comparison of dimension scores for Dimension 3: Managed information flow ANOVA: Registers, F=21.852, p<.0001.*

potentially to signal the introduction of an instruction given to the caller or customer (e.g., "*Ok, sir, let's send this order to customer service and wait for their response ...* ").

Both Maritime and Aviation texts fall on the negative side of Dim 3, clearly suggesting that the management of the flow of information is not encouraged and not necessarily expected in these two settings, compared to other telephone-based interactions. In call centers, the use of the positive features of Dim 3 could be related to common conventions in customer service such as establishing rapport, avoiding "dead air," as well as backchanneling to show attentiveness and focus on the customer in the transactions. To this end, Filipino and Indian call agents undergo skills training in phone-handling, and some of the topics covered in many training sessions include backchanneling and providing confirmatory responses to control the flow of transactions. Some researchers have noted that Filipino agents tend to be quiet during callers' turns which may suggest to the callers limited engagement or low level of interest (Friginal, 2009; Peltzman & Fishburn, 2006). Because of this awareness during training, it is possible that agents consciously backchannel in their turns. Interlocutors in Maritime and Aviation texts do not actually need explicit backchanneling due to the common conventions in radiotelephony to acknowledge speaker turn endings and confirm comprehension, as if speakers are using one-way channels (e.g., *roger, copy, copy that*). This phenomenon came from a tradition of sharing a tactical single-channel radio signal even when most interactions, especially in aviation, have long since relied on advanced multi-channel communication technology and interactions.

Call Home and Switchboard discussions are not task-based, with limited imperative *let's/let us,* but communicative markers such as *ok, actually, basically, exactly*, and *anyway* are also frequent in speakers' turns. In managing the flow of information and trying to control turns in telephone-based talk, it appears that speakers are addressing three unique objectives: (1) direct management, that is, avoiding dead air, confirming the message, initiating the turn; (2) indirect management though speaker mannerisms and speech patterns; and (3) making use of the positive features to supplement fillers to "buy thinking time" before a response (Friginal, 2008, 2009).

Text Sample 6.7 illustrates a call-taker's use of the positive features of Dim 3, while Text Sample 6.8 shows similar patterns from two speakers in Call Home. *Ok, anyway, let's,* and *well* co-occur with adverbials *actually, supposedly, exactly*, and *basically* in the call-taker's turns. *Ok* and other discourse particles often start the call-takers' turns, and sometimes are used together to mark the beginning of utterances. In several instances, *ok* is also used to signal transitions or turn endings. Adverbials often belong to different semantic categories with different discourse functions. In this context, stand-alone adverbial *exactly* is used as a direct, confirmatory response, while stance adverbial *actually* implies verification of information (e.g., *"..actually June 30"; "I actually checked your.."*). Note again that in Maritime and Aviation texts, these instances are rare.

Text Sample 6.7. Purchasing transaction (Dim Score = 4.318)

    Caller: [long pause] Uhm one of them was I believe was on I believe was on the 25th of June

    Call-taker: **Ok?**

    Caller: Two of them was on the 25th and one of them was on the 21st of June

    Call-taker: **Ok, let's** just go ahead and check [long pause] **ok** [hold 22 seconds] the other one I believe was on the two you have **actually** won three recruits right?

    Caller: Yes

    Call-taker: **Ok** you have three recruits so **let** me just check [long pause] **ok** so it is here that since you recruited them just last 25th they supposedly [long pause] **ok let** me just go ahead and check on this, I'll call you back because I **actually** checked your [XX Account] and that coupon is not loaded in your [XX Account] **ok?**

    Caller: [unclear] I don't see it there

    Call-taker: Yes, yes and uh you know the start is **actually** June 30 **well** but **anyway** you have until the end of this month to redeem this coupon **basically, so whatever, let** me just go ahead and check why the coupon is not loaded

Text Sample 6.8. Call home—travel schedules (Dim Score = 3.213)

    Speaker 1: **Right, right.**

    Speaker 2: **Right.**

    Speaker 1: But **anyway,** they they live in New York City in Queens and they're **really** nice so if you get stuck like you know I could give you their number or something.

    Speaker 2: **uh-huh ... yeah well** no I hopefully it will be okay

    Speaker 1: **yeah.**

    Speaker 2: um what was I going to say um ah so **anyway** uh I'm **basically** what I'm doing I'm I'll be **just** like only four days in Buffalo.

    Speaker 1: **uh-huh.**

    Speaker 2: And, of them, in fact, only the Grandma Ruth and Grandma Henning will only be there for three.

Speaker 1:	**Really?**
Speaker 2:	The last the last day I'll be spending with my friend, Cathy.
Speaker 1:	**uh-huh**
Speaker 2:	And then I'm going to Johnny and I think we have
Speaker 1:	When are you getting to Johnny's?
Speaker 2:	I'm flying in on the twenty-eighth.
Speaker 1:	**oh,** the I'll **probably just** be leaving Michigan then.
Speaker 2:	By ya
Speaker 1:	Like to drive to South Carolina.
Speaker 2:	**uh-huh.**
Speaker 1:	I

6.5. Discussion and concluding remarks

The analysis of spoken, radio/telephone-based interactions using the MDA framework reveals several interesting similarities and differences across corpora. The three dimensions (*addressee-focused, polite, and elaborated information vs. involved and simplified narrative; planned, procedural talk;* and *managed information flow*) illustrate the primary patterns of lexico-syntactic features of business interactions and how these may be used in Aviation and Maritime texts. Pilot-ATC utterances, in summary, are different in linguistic composition from business call centers, even with their clear contextual parallels with agents-callers in the medium and also the various functions of their discourse features. Aviation and Maritime texts are found to be very similar in linguistic patterning across the three dimensions. The analyses here focus on broader register comparisons, and the next step is to further pursue comparisons by differentiating the texts from the corpora according to speaker roles (e.g., agents vs. callers, pilots vs. ATCs) and language background (e.g., English NES vs. NNES). The information coming from ATCs is primarily planned and procedural, often similar to how agents frame their utterances, but the similarities end there. In call centers, agents constantly manage and monitor their utterances, highly focused on establishing rapport and working together (e.g., *let's start with the third part; we'll have to change your password*). Agents have the data to share as well as the instructions and procedures to resolve an issue, just like ATCs, but their ways of delivering them to their interlocutors are very different. The callers (and pilots) expect these procedures and information, for the most part, unless they call to complain or express dissatisfaction. Pilots rely on ATCs for specific information and instructions, but there are fewer unexpected questions or concerns, especially outside of emergency situations.

What then is the relevance of these comparative results, especially for aviation professionals? In Dims 1 and 3, in the spirit of service and personalization of support, agents use politeness markers frequently, engage the callers by giving sufficient or detailed information and explanation, and use discourse markers to monitor the flow of conversation. These patterns are not necessarily encouraged—or even necessary to the task at hand—in aviation talk and *plain language*, and clearly ATCs (especially US-based ATCs) do not make use of these patterns frequently in their utterances. However, given the intercultural and global nature of aviation discourse, other ways of delivering instructional and task-focused language may have to be examined more closely. There are several texts, especially some intended for use with non-native English-speaking pilots, in which personalized and polite support appears to be recommended and preferred in consideration of the characteristics of these NNES, some of which have been identified above, in the interest of accurate and complete communication within the airline industry.

The question, of course, follows as to whether or not politeness features and respectful language are really necessary in aviation rather than what some in the industry might describe as more task-focused or "direct," "efficient," or "no frills" communication. In Text Sample 6.9 below, an instance of the international pilot (Etihad503—ETD503) communicating with an ATC from New York (JFK Airport) provides an opportunity to consider this question. The pilot in the recorded interaction actually used high-frequency features of Dims 1 and 3 (e.g., *sir, please, thank you, okay*) and explicitly and politely asked the ATC to be polite (line 19 below). The audio and transcription of this interaction are publicly available at: https://www.youtube.com/watch?v=ZWOOKQIEe5s

Not captured in the transcript are the tone, style (accent), and related suprasegmental features of the ATC's questions, which clearly affected the pilot's perception of the interaction as impolite with unfortunate negative consequences.

Text Sample 6.9 Be polite with me (Dim 1 Score = 1.54)

Description: JFK ATC talking with multiple pilots on airport taxiway for routine landing procedures.

1. JFK GND: Echo Tango Delta 503, where are you parked?
2. ETD503: Bravo 28, **sir**
3. JFK GND: Not taxiway, the letter
4. ETD503: **Oh negative sir,** we are on 22R holding short of Foxtrot
5. JFK GND: What taxiway do you enter the ramp?
6. ETD503: **Okay sir,** we just exit the runway and we're holding short of Foxtrot on 22R.

7.	JFK GND:	You're not listening to what I'm asking you. What taxiway do you enter the ramp?
8.	ETD503:	I'm not on the ramp yet **sir.**
9.	JFK GND:	What taxiway do you enter the ramp? Tell me. What letter?
10.	ETD503:	**Okay,** we can enter at Kilo for Etihad 503
11.	JFK GND:	That's what I need to get out of you, we talked like six times. Straight ahead and hold short of Hotel, sir.
12.	ETD503:	Straight ahead hold short of Hotel, roger.
13.	JFK GND:	Asiana 222, turn right here. Turn left on to 22R, and hold short of Juliet.
14.	JFK GND:	Asiana 222?
15.	AAR222:	Go ahead?
16.	JFK GND:	Turn right, left on the 22R. Hold short of Juliet.
17.	AAR222:	**Okay,** right turn, then 22R, holding short Juliet, Asiana222.
18.	JFK GND:	ETD503 follow Asiana222, hold short of Juliet on the runway.
19.	ETD503:	**Uh, yes sir, we'll follow Asiana, but next time I would like you to be polite with me. Thank you.**
20.	JFK GND:	Okay, but if I got to talk to you six times, and I got all other people I got to talk to, and you don't understand what I'm saying.
21.	ETD503:	… (unclear) nice day, **be polite with me. Alright?**
22.	JFK GND:	Uh, you're impolite with me
23.	ETD503:	I'll make a report
24.	JFK GND:	Go ahead.

In further understanding intercultural pilot-ATC communication in aviation, it is relevant to conduct additional and follow-up qualitative research, considering how pilots react to ATC questions, direct instructions, and how routine and emergency discourses are facilitated. How are these similar tasks performed in other locations outside the United States? What are common language-related communication problems or concerns by US pilots (who are native English speakers) in these other locations? These questions are relevant to the linguistic discussions in this chapter and will further contribute to descriptions of the *lingua franca* of global aviation. Results from these types of analyses may further indicate the influence of cultural variables in the linguistic choices of speakers in this domain. In addition, issues about pronunciation and oral speech production variables (which are not considered in this chapter) are certainly important subjects for future research. Clearly, features of verbal communication not encoded in the transcribed texts of transactions are not captured by corpus tools. MDA and corpus-based approaches to spoken discourse are still limited in accounting for phonetic factors and oral production circumstances primarily because orthographic text transcriptions are mostly limited to lexical and structural features (Baker, 2006).

Thanks to **Aline Pacheco** and **Joao Cavallet** for sharing a copy of the Corpus of Pilot and ATC Communication (CORPAC).

PART THREE

Pedagogy

7

Aviation English Pedagogy: Contexts and Settings

7.1. Introduction: The landscape of aviation English

Aviation English training worldwide is urgent, high-stakes, and critical to business and economic stability in any aviation environment, especially where participants do not share a common first language. Although current global businesses and migration depend greatly on the ability to travel from nation to nation safely and efficiently, the careful, thorough, and systematic attention which should be given to the development and implementation of aviation English teaching and learning infrastructure and outcomes to accomplish this objective vary across locations. Ineffective, inefficient, and unreliable programs are, unfortunately, common in the industry due to the largely unregulated implementation of ICAO Language Proficiency Requirements (LPRs) discussed previously in this book. The challenges of providing programs which comply fully with Best Practices for language training are encountered throughout the industry and have created a market which is quite limited and primarily proprietary or commercial in nature (Borowska, 2017). Companies and organizations looking for an aviation English program do not currently have many options, particularly for training programs which are informed by and based upon sound academic theory and practices and innovative ESL/EFL or ESP/EOP models. The Standards and Recommended Practices, including a rating scale and a set of holistic descriptors, provided by ICAO were undeniably an important step in advancing the teaching of aviation communication, but without clearly operationalizing these descriptors into standardized curricula appropriate for various professions across levels, aviation English continues to experience major limitations and a range of continuing obstacles (Campbell-Laird, 2006).

When ICAO instituted its LPRs, it did so, for the most part, into an industry that lacked the essential infrastructure to operationalize them. A typical English-medium institution, for example, may require specific test results such as those from the Test of English as a Foreign Language (TOEFL) or International English Language Testing System (IELTS) from its prospective international students, with support from a robust language learning program monitored by accreditation bodies and staffed by qualified language instructors. In contrast, the UN's aviation English policy (i.e., ICAO LPRs) does not have a clear institutional structure, an implementing agency, and/or established resources. Globally consistent standardized tests, trained test raters and aviation English instructors, accrediting bodies, a range of culture-specific teaching and learning materials, and sufficient research in the field of Language as a Human Factor are inadequate and developed primarily

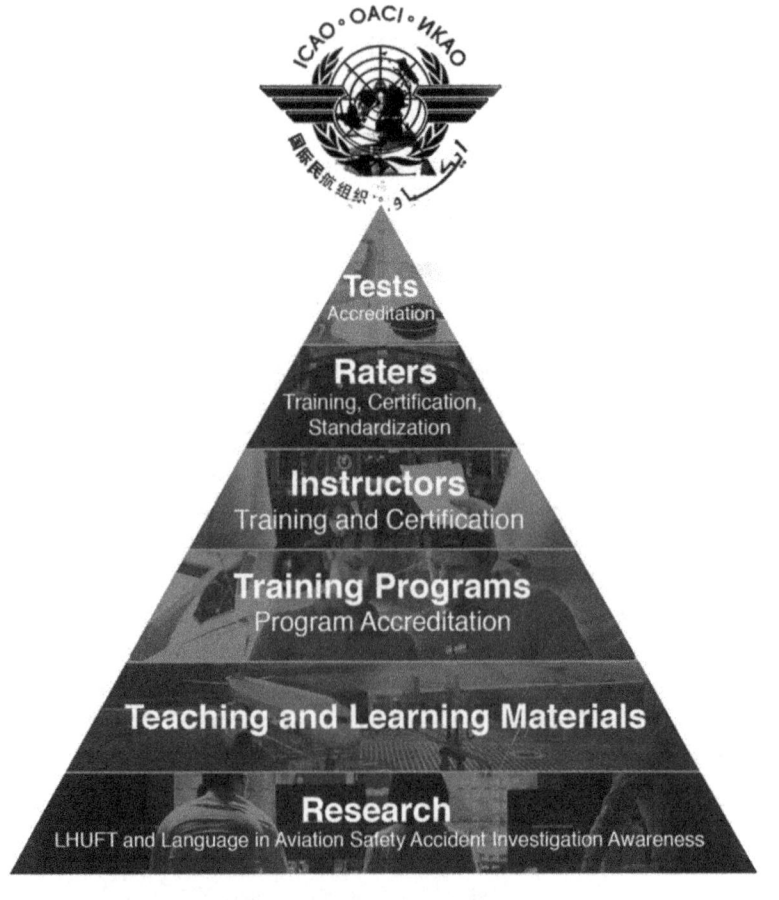

FIGURE 7.1 *Pyramid model of required infrastructure to comply with ICAO LPRs (developed by Mathews for this book and for Embry-Riddle).*

in isolation by various institutions. The necessary organizational support to ensure that all ICAO Member States can comply fully with ICAO LPRs has been documented and often discussed; however, actual capital to build a dependable infrastructure is still lacking.

As more applied linguists and TESOL professionals have been gradually integrated into aviation English pedagogy over the past ten years, an increasing focus on the training of non-native English-speaking (NNES) pilots, air traffic controllers (ATCs), and maintenance staff has been given some level of priority. As mentioned in the previous section, there is also an increasing number of collaborative research initiatives in various domains and foci in the industry. It is important to note, however, that to effect the most important and beneficial improvements in safety and success in day-to-day operations in the aviation industry, the development and implementation of professional, thoughtfully designed, research-based, university-affiliated language training programs is essential. Such programs could produce consistent and globally focused testing, teaching, and learning materials. The pyramid model in Figure 7.1 identifies and shows the interrelationships of these components of an ideal infrastructure designed to fully implement effective compliance with ICAO LPRs.

7.1.1. What are we teaching?

ICAO Doc. 9835 (2010) defines the field covered by the term *aviation language* quite broadly, including:

> " ... all of the language uses of many different professions (engineers, technicians, commercial staff, flight crews, etc.) within the aviation domain, which itself includes specializations such as aircraft construction, aircraft maintenance, aircraft operations, air traffic control, regulation, airport activities, passenger care, and flight crew operations." (3.2.6)

With regard to ICAO LPRs, however, aviation language applies specifically to aeronautical radiotelephony communications, identified as a specialized subcategory corresponding to a limited portion of the language of only two aviation professions—controllers and flight crews, especially their use of ICAO standardized phraseology and plain language. This definition of aviation language may cause some confusion, then, as to what "aviation English" is and should cover regarding language teaching, training, and testing. Although ICAO LPRs are only for professional pilots and controllers, aviation language extends to other various specializations in the larger field of aviation, as shown above.

Borowska (2017) suggests that the term a*viation English* be considered as the primary construct for all aviation domains, while the terms *Aeronautical Language* and *Aeronautical English* be designated to refer specifically to radiotelephony (RTF), including both standardized phraseology and aeronautical plain language. RTF and aviation English are certainly not interchangeable here; RTF is an integral part of aviation English, but there exist many English-medium communicative tasks in aviation which do not take place over the radio. This distinction is of significant importance in pedagogy because course contexts should consider the exact domain of the target language since that influences and directs the nature of instruction and the target goals.

The teaching and learning of aviation English has several distinctive characteristics distinguishing it from the broader teaching and learning of English in other registers. In particular, aviation English directly addresses safety components for flight crews and their passengers as well as relevant peripheral benefits to airlines' and nations' financial well-being. In addition, the language of aviation is unique in that it employs a specific sets of vocabulary, phrases, and speech acts, and it places the greatest importance on achieving operational efficiency rather than linguistic correctness or grammatical correctness (Kovtun & Simoncini, 2014). In this chapter, for the purpose of discussing pedagogy in the world of aviation English, the focus is on not only professional pilots and ATCs, but also on all aviation personnel who need English to communicate and successfully complete work-related tasks.

English is most assuredly the *lingua franca* of civil aviation, even in educational settings, sanctioning communication in major, especially official, domains. It is clear, however, that most of today's aviation operational personnel do not have English as their first language (Emery, 2015). Pilot-controller communication may receive the most attention, but almost anyone working in the aviation industry globally will be expected to produce or respond to English language during their career. English is used when pilots communicate in the flight deck or during briefings; when cabin crew use English to communicate with passengers, pilots, and each other; when aircraft engineers write technical reports detailing complex systems and instructions; when aviation maintenance technicians read Boeing or Airbus manuals to repair an engine; when airline businessmen form and interact within global partnerships; and so on. Aviation English, therefore, encompasses many contexts within the aviation industry. Communication within all of these contexts impacts safety in one way or another, and all deserve thoughtful attention in the teaching and learning of the specific type of English necessary to perform a job in each context.

7.2. The aviation English learner

Aviation English training occurs in hundreds of locations, with thousands of learners, from teenage pilot cadets in the United States who have yet to control an aircraft, to ATCs with decades of experience at busy airports in Thailand. Language training occurs intensively in various environments, covering spans from weeks to months for many hours a day, as well as in shorter episodes of just a few hours a few times a year. Although these training models differ dramatically, administrators and learners often expect equivalent learner proficiency results from dramatically different pedagogical scenarios. Airlines; flight schools; air navigation service providers (ANSPs, or any entity which provides air navigation services); civil aviation authorities (CAAs); maintenance, repair, and overhaul (MRO) facilities; or even aeronautical universities may provide training for aviation personnel in various capacities, with a wide range of resources, instructor qualifications, and time devoted to training. The array of learner demographics is perhaps one of the most diverse within a single industry, which presents its own set of unique challenges for which curriculum designers and instructors must find solutions.

One of the key challenges confronting the industry is a result of the limited scope of ICAO LPRs, which, again, focus only on pilot-controller communication and do not address the language proficiency required for safe and efficient flight training, maintenance, or even cabin crew operations. Despite a lack of formal regulation in these areas from ICAO,[1] almost all personnel must still be able to utilize English in the course of their work, especially if they are to work in any capacity internationally without posing potential, serious safety risks to themselves and others. The following sections focus on a discussion of aviation English teaching/training contexts or settings, from the needs of commercial airline pilots, to maintenance staff, and to target subject areas like flight training. Although not an exhaustive list, the variety of contexts identified below illustrate the many facets of aviation training which require proficiency with the English language for safe and productive operations.

7.3. English for commercial airline pilots

7.3.1. Context

Pilots have primary responsibility, of course, for the safe transport of passengers and crew all over the world. In order to do so, pilots must effectively communicate not only with ATCs, but also with other members of the flight deck, pilots in the vicinity, maintenance technicians, ground staff, cabin crew members, and even passengers. Pilots rely on communication with ATCs for information about altitude, direction, speed, weather, take-off and landing instructions, and other applicable traffic information (Estival, Farris, & Molesworth, 2016). While an airline's safety record and reputation are built on many factors, pilot communications to ATCs, including their intelligibility, effectiveness, and proficiency, are not only primary determinants of safety, but also influence the external perception of the airline. Controllers and other pilots communicating with an airline captain cannot know the pilot's total number of flight experience hours, but the judgment of the pilot's proficiency will occur nevertheless, based largely on the real-time communication that occurs and will, understandably and fairly, be a reflection on the airline. Additionally and most importantly, all airlines should be committed to achieving and maintaining the highest possible international standards of operational safety, and successful English communication is certainly instrumental in accomplishing this.

7.3.2. Language training

Traditionally, the main focus of commercial airline pilot aviation English training is on speaking and listening skills. However, although not regulated by ICAO, English reading and writing skills are also necessary in the operations of commercial airlines. For example, written language appears on flight deck control labels, display screens, dispatch paperwork, navigation charts, checklists, operation manuals, and even notes on ATC clearances (Holder, Hutchins, & Nomura, 2002). The complete lack of attention to ensuring proficiency in these language skills for non-native speakers of English as a component of training programs can certainly pose a threat to safety and inefficiency and should be addressed in language courses, even if only peripherally. The aviation English trainer in this context typically works with highly experienced professionals whose time is valuable and limited. The time spent in the course must, therefore, be optimally utilized. It is also likely that, in this context, the student has more operational knowledge and RTF experience than the trainer (ICAO, 2009). Furthermore, specific local training variables must be considered such as the dominant language of the workplace, availability and commitment of learners/students, and the recurrent testing procedures in place (Mell, 2004).

7.4. English for air traffic controllers

7.4.1. Context

The primary job of an ATC is to ensure that aircraft are correctly and safely positioned and maneuvered throughout the various stages of flight, especially during take-offs and landings, by issuing instructions to pilots. ATCs are responsible for communicating with pilots throughout a flight, beginning with ground control before the plane takes off. Controllers also coordinate with one another as an aircraft is handed off from one controller to the next during the course of a flight. These duties are performed while sitting close together in front of computer screens and often while communicating with multiple pilots during short periods of time (Estival, Farris, & Molesworth, 2016; Sullivan & Girginer, 2002). The work of an ATC is universally regarded as notoriously taxing due to the extremely heavy, fast-paced workload and large volumes of air traffic to manage in major global hubs. Appendix B of ICAO Document 9835 (2010) provides a list of over one hundred language tasks of an ATC, including categories such as managing air traffic sequences, controlling aircraft ground movement, performing situation monitoring, and resolving aircraft conflict situations.

The American Airlines Flight 965 accident near Cali, Colombia, discussed in Chapter 1, during which the controller was unable to formulate questions to clarify the flight crew's seemingly illogical statements concerning the plane's location (Flight Safety Foundation, 1998), dramatically demonstrates the critical need for highly skilled ATCs. This accident certainly points to the need for English language proficiency that extends beyond routine, standardized phraseology. Controllers need to be able to respond to questionable or unclear transmissions from the flight deck, identify and clarify ambiguity, and then promptly provide flight crews with clear instructions to address these situations.

7.4.2. Language training

According to Uplinger (1997), "teaching and testing knowledge of ATC terminology with lists of terms turns controllers into parrots, who are handicapped in unusual or stressful ATC situations, rather than skilled users of English who can apply the language in a range of contexts" (p. 3). While some recurrent training in phraseology may be useful for ATCs in this context, to build and maintain their general, plain English proficiency, a focus on more sophisticated and complex topics relevant to a range of ATC tasks should be used and directly incorporated into the training environment (Uplinger, 1997).

As with pilots, in-service ATCs are provided with limited training time despite steadily increasing air traffic complications around the world. At one ANSP in which all staff are NNES, recurrent English language training is limited to two- or three-day classes, four times a year. In such a situation, it is vital that the curriculum takes full advantage of each minute in the classroom. Kay (2015) suggests that a successful English-for-ATC curriculum should include engaging content, be based on real-world situations, maximize peer to peer interaction, appear useful to learners, and allow learners with a range of levels to participate. While these criteria may seem obvious to the experienced ESL professional, the realities of many existing training programs, unfortunately, do not reflect these basic principles.

7.5. English for pilots and controllers

Although ICAO LPRs do not distinguish descriptors or criteria for pilots and controllers and the aim for both is to achieve, minimally, Operational Level 4, the training provided should, nevertheless, be tailored for these professions separately. Kovtun and Simoncini (2014) suggest that pilot training should explore a wider range of operational situations, registers of communication, and interlocutors, since pilots are required to communicate with, for example, other flight crew members and maintenance technicians. On the other hand, air traffic control training may focus more on operational content such as the language of airspace and interaction with emergency services, for example, since ATCs are more likely to need this subset of language.

7.5.1. Context: Pilot-controller communication

Miscommunications between pilots and ATCs undoubtedly occur quite often, but fortunately, they do not often result in accidents or serious incidents. However, the potential for miscommunications which cause financial loss, wasted time, or even fatalities is always very real. English language proficiency and other factors such as workload, audio signal quality, and use of standardized phraseology are all possible contributors to communication errors (Molesworth & Estival, 2015). Among these factors, pilots typically perceive understanding other pilots as the most challenging, but factors such as the number of items in a single transmission also rank high (Estival & Molesworth, 2009).

Communicative functions. ICAO Document 9835 (2010) identifies four communicative functions of pilot-controller communication, including (1) triggering actions, (2) sharing information, (3) managing the pilot-controller relationship, and (4) managing the dialogue. *Triggering actions* is described as "the core role of pilot-controller communication" (3.4.7), categorized by speech acts that trigger specific responses, including giving orders, giving advice, requesting actions, or requesting permission. *Sharing information* is the necessary counterpart to the above speech acts in order to effect communication "because appropriate actions can be triggered only if the pilot and controller share sufficient information about the current situation" (3.4.7). Examples of sharing information include stating or asking about one's intentions, availability, preferences, actions (past, present, or future), or readiness. Greetings, complaints, apologies, and expressions of thanks, satisfaction, or concern are communicative tools used to manage the relationship between pilot and controller. Finally, to ensure successful management of the dialogue, speech acts such as paraphrases, comprehension checks, repairs, clarifications, or confirmations may be used. In order to successfully participate in this communication, participants need shared familiarity with these functions, events, domains, and tasks in the aviation context, as well as specific pragmatic and strategic skills necessary to manage them. In order to avoid miscommunication, focus on these speech acts using content relevant to their profession results in a training experience which has validity and value because it is skill-based and corresponds to trainees' real-world context. These speech acts are important components of any language training curriculum for professional pilots or controllers.

ICAO holistic descriptors. ICAO provides five holistic descriptors as a frame of reference for training for pilot-controller communication. The descriptors are outlined in Table 7.1.

The first descriptor refers to the language of RTF, and has a few defining characteristics that present not only challenges to learn and perform, but also challenges to teach. Radiotelephony contains both the prescribed set of standardized phraseology found in *ICAO Document 4444: Procedures for Air Navigation Services—Air Traffic Management and Document 9432: Manual of Radiotelephony*, and what is termed "plain English" in an aviation context. Phraseology is not only a set of terms and phrases, but also includes rules concerning pronunciation and syntax which differ from natural language (Estival, Farris, & Molesworth, 2016). Plain language is used when phraseology does not suffice, and a form of code-switching occurs

Table 7.1 ICAO language proficiency requirements holistic descriptors

In addition to ICAO Rating scale (six levels with six criteria), as outlined in ICAO holistic descriptors, ICAO also requires that Aviation English tests used for licensing purposes need to be able to assess how well pilots or controllers can:

1. Comprehend information and communicate effectively in voice-only telephone and (radiotelephone) and face-to-face situations;
2. Communicate on common, concrete, and work-related topics with accuracy and clarity;
3. Use appropriate communication strategies to exchange messages and to recognize and resolve misunderstanding (e.g., to check, confirm, or clarify information) in work-related contexts;
4. Handle successfully and with relative ease the linguistic challenges presented by an unexpected turn of events that occurs in the context of flight/work operations in air-ground situations or communicative task with which they are otherwise familiar; and
5. Use and comprehend accents which are intelligible to the wider aeronautical community.

when users must switch from phraseology to plain language. In addition, RTF is done in a non-visual, voice-only format, lacking the paralinguistic features present in face-to-face communication, such as facial cues and body language (Bullock, 2015; ICAO, 2010; Petrashchuk, 2015). Sound quality and transmission can be another challenge, as both may be compromised by busy radio traffic, interference, or faulty equipment. As precision and conciseness are very important, RTF places an emphasis on content rather than linguistic form, accomplished by the use of formulaic sentence fragments rather than grammatically complete sentences. Often, extraneous function words such as determiners and auxiliary verbs are deleted. Regarding phraseology specifically, Uplinger (1997) recommends that it not be taught to beginners, but rather to students who have an advanced knowledge of general English. The high level of functional proficiency needed to understand RTF in a second language is due to not only its specificity, but also its infrequency in general language. It is in fact markedly different than natural language, in terms of syntax or lexicon, and is, on its own, much like a foreign language (Campbell-Laird, 2006; Mitsutomi & O'Brien, n.d.). Knowledge of essential general English must, therefore, precede the learning of this highly complex language.

The second descriptor states that pilots and controllers should be able to communicate effectively about any matter related to their professions and work-related responsibilities, in the air and on the ground, using phraseology and plain language. Notably, "plain language" in the aviation context differs from other domains where the intention is to communicate specialized language intelligibly to non-specialized individuals, making use of simplification and avoiding technical jargon. In contrast, aviation plain language is still spoken only within the aviation environment among insiders, including technical terminology (Estival, Farris, & Molesworth, 2016). Plain language must, therefore, be targeted during instruction, as pilots and controllers may not have opportunities outside training to develop this specific language domain applied to the aviation-specific context. The intent is not to replicate natural conversation in English, "but the use of simple English following as much as possible the guidelines provided by the phraseology and obeying the constraints of aviation communication" (p. 38). While this descriptor was written specifically for pilot-controller communication, it is also a helpful guideline for curriculum designers for other contexts, such as maintenance or cabin crew, and is applicable generally for an aviation content-based program in all contexts.

The third descriptor refers to the ability to strategically and accurately utilize language skills to reconcile misunderstandings that arise in the course of carrying out interactive work responsibilities. Small miscommunications in aviation can have fatal consequences. In addition, although pilots and controllers have prescribed phraseologies with standardized syntax and procedures, the fourth descriptor stipulates that they must also have the necessary linguistic flexibility to be able to communicate effectively to manage unfamiliar and/or unexpected events that may arise. Situations can change abruptly, and pilots and controllers, working together, must be able to react quickly and appropriately.

Lastly, inasmuch as eliminating an accent is not a specified or reasonable goal, the fifth descriptor stipulates that to the extent that accents are present and used in communication, they must be intelligible so as not to impede accurate and complete understanding and comprehension, an essential requisite for all NNES training curricula (Murphy, 2014). ICAO suggests the model in the UK of screening applicants and judging marked accent by intelligibility only to those most familiar with the dialect. ICAO makes clear that Operational Level 4 does not target or require native-like pronunciation as a goal, but evaluates pronunciation operationally based upon its apparent effects on successful communication and comprehension.

7.5.2. Language training

Actual videos and recordings aligned with these ICAO descriptors are useful teaching tools if they are engaging, relevant, of good sound quality, appropriate length, and level-appropriate. The instructor must then determine themes, vocabulary, communication issues, and discussion topics that are just beyond the learners' current language level or realm of experience. Lessons must be designed around specific learning objectives, such as identifying causes of misunderstanding, listening for key ideas when aircraft report issues, or revising language strategies to clarify key information (Kay, 2015). Classroom activities such as information gaps, gist listening, paraphrasing activities, and role plays can then be used to accomplish the particular learning objectives. The high degree of authenticity of the "real world" training materials, both the content and the delivery itself, followed up with high-quality language competency/proficiency testing, ensures real and meaningful language improvement that enables learners to communicate effectively in real operational situations (Kay, 2017).

Commercial airline pilots and professional ATCs may approach the learning of the English language with very different attitudes. On the one hand, they see English skills as useful in their jobs and appreciate the importance of strong language proficiency in their work. However, being forced to take an ICAO language proficiency exam—particularly to those with decades of experience without any adverse incidents resulting from inadequate language proficiency, as far as they're aware—may be negatively regarded and even resisted. For this reason, course designers *must* do everything possible in the design and implementation of the training to ensure that the content in aviation English courses and ICAO language proficiency exams has the greatest potential possible to be perceived by the learners and test-takers as valid, accurate, and interesting. It should be noted, in this regard, that testing which closely matches real-world communication as much as possible has greater validity and, consequently, more positive responses from anxious or skeptical students; when real-world target language situations, training, and testing all align, learner response and results are more likely to be positive (Bullock, 2017; Kay, 2017).

The quality of a language test can have a direct impact on the effectiveness of language training programs, influencing proficiency development by affecting student attitudes toward learning (Kay, 2017). Many language trainers wish to ignore the realities and importance of aviation language testing, perhaps because of the pervasive and onerous perception of testing and evaluation, and focus, instead, only on adequate training. Obviously, well-designed, purposed instruction has great merit and is essential, but it is also possible to utilize a testing instrument—perhaps better described as an evaluation strategy—that is so well or creatively designed with a real-world focus and so well integrated into the training program itself that the testing-training dichotomy is imperceptible, and the necessary and valuable evaluation/testing aspect of the training is not offensive to the learners.

7.6. English for cabin crew

7.6.1. Context

Cabin crew are, of course, essential for the safety and comfort of passengers, and also for several other major safety and hospitality reasons, including the conveying of standard briefing information and providing instructions during an emergency (Wegler, 2015). The demand for cabin crew worldwide is growing exponentially, with a projected need of about 890,000 new cabin crew by 2037, with 321,000 in the Asia-Pacific region alone. This major upsurge in NNES cabin crew also increases the need for focused English for Cabin Crew training development. The language training needs of cabin crew differ significantly from those of pilots and controllers; however, there is some content overlap of operational aviation knowledge which can be utilized for crew members as well as pilots and controllers by focusing on the particular roles and responsibilities unique to each employee group during routine or non-routine aviation events in which all employees are involved.

7.6.2. Language training

Arinik (2015) demonstrates the use of an informational video designed for pilots on refueling with passengers on board to train cabin crew members in communicating effectively and conveying accurate information. By pausing the video at specific points and using images/events as prompts, students are asked to perform role plays of possible situations in the scenario in which cabin crew would need to communicate. For example, the video explains the importance of ensuring that equipment is not obstructing emergency exits during this procedure. Learners are asked to report to the cabin chief that there is an obstruction in the emergency exit. The cabin chief is then asked to inform the captain. Following this sequence, the video proceeds to instruct that any detected fuel smell means that refueling must stop immediately. Learners then communicate this situation is occurring from a passenger to a cabin crew member, a cabin crew member to a flight crew member, and finally a flight crew member to ground staff, with reminders for heightened attention to language and word choice differences potentially affecting comprehension at each step of the scenario as it proceeds.

7.7. English for ground staff

7.7.1. Context

As is sometimes the case with cabin crew, airport ground staff including bus drivers, caterers, security guards, airport information desk staff, and ground handlers are often underrepresented in conversation, research, and development in ESP, perhaps due to the incorrect assumption that staff in these positions are not routinely involved in what might be "life or death" English communication scenarios. However, depending upon sometimes unusual and unexpected, potentially emergent situations and unpredictable factors and sequences of events in such situations, the lack of meaningful English language proficiency enabling people in these positions to communicate effectively may, in fact, impact safety (Cutting, 2012). Ground staff are aware of their English language needs, as evidenced in the results of a survey given to thirty ground staff at Phuket International Airport in Thailand (Phithakphongphan, 2016). Respondents revealed a desire to improve their listening and speaking skills, which they felt were most relevant to their jobs, and also reported believing that improving their English proficiency was necessary to achieve promotions. Results from another study of airport information staff found that respondents aspired to gain "effective on-the-spot communication with passengers" (Ting, 2010, p. 6).

Examples of near misses due to miscommunications with ground handlers include a plane nearly going off the end of the runway due to a miscommunication between a captain and ground staff about a tow bar (NASA, 2005, as cited in Cutting, 2012). According to the ground staff's communication, the captain believed the tow bar was already connected and he released the brakes, expecting that the aircraft would not move. English language proficiency of ground staff, although not regulated, should still be considered as necessary to airports, not only for safety reasons, but also for economic reasons, as ground staff are often the face of customer service for passengers and should be able to use English with appropriate markers of politeness, clarity, and *informativeness* (Cutting, 2012). Airport information desk staff, for example, need to give information about check-in, baggage, and directions, as well as provide instructions for passengers with special needs (Ting, 2010). Some of the busiest airports in the world, for example, Beijing Capital International Airport, the main airport in Beijing, have responded to very low passenger satisfaction surveys (ranked 62nd; see Annual Conference of Civil Aviation Administration of China survey results, 2008, as cited in Ting, 2010) by working to improve staff training, including their English proficiency. Initiatives of this nature, with much specificity, are uncommon, however. Included under the umbrella of "airport ground staff" are many occupations with various English language abilities and needs. The necessary allocation of money and time to create customized, comprehensive, effective training courses is often unacceptable to industry decision-makers, and individuals are, instead, given more generic training or none at all.

7.7.2. Language training

When designing curricular materials for this demographic of aviation personnel, the amount of time afforded and the best environmental context for learning must be considered. The ELSY Project (*ELaboration d'un SYllabus multimédia aéroportuaire pour les jeunes sans emploi et peu qualifies*) in Charles de Gaulle Airport in Paris was aimed at providing employment for a second-generation immigrant population who lacked the English language skills to obtain jobs as airport ground staff. The project included the development of learning materials for airport ground staff with the cautionary note that the materials would most likely be accessed from home or in spare moments between jobs, and would likely be disregarded by staff if they were found to be too grammar-focused (Cutting, 2012). A discourse analysis was done to identify the most frequent functions of ground staff communication and to determine which grammatical features best enabled staff to perform these functions. The results of this analysis then informed the development of teaching materials which highlighted critical information and contexts useful in the actual performance of the job, thereby eliminating extraneous materials and unnecessary linguistic tasks in the training program.

7.8. English for maintenance

7.8.1. Context

In recent years, "English for Maintenance" (i.e., for maintenance writers, technicians, and staff) has been gaining more traction in the industry, but a scarcity of research and the development of effective curricula for this employee group still exists. Professional maintenance materials are almost always written in English, as discussed in Chapter 5, despite the reality that nearly 80 percent of their users are NNES. A study spanning four continents, 113 airlines, and including 941 aviation maintenance workers concluded that there are major training needs, not only in the area of reading comprehension of personnel as presented in Chapter 5, but also in the development of consistent, readable, and systematic technical texts (Drury, Ma, & Marin, 2005). What makes these results interesting—and potentially alarming—is that all maintenance workers across locations are expected to comprehend the same English-language maintenance texts.

Translating maintenance manuals and instructions is expensive and fraught with potential inaccuracy risks, and doing so is actually less common than the typical lay observer might expect—or hope for. In addition, although some manuals are translated into the national or dominant language of a territory, there are no regulations or standards to require validation of the accuracy and completeness of any technical document in the industry (Transport Workers Union, 2018). In an industry where precision is critically important, the risks of poor translation and the requisite cost of this "extraneous" task may be too high to take. The risks of not doing so, with accuracy, however, might be higher.

Aircraft maintenance personnel are responsible for making decisions and for executing tasks according to the information provided in these manuals. The results of these decisions and tasks unfortunately end up frequently being reported as "human error" or "judgment error" in the documentation of technical malfunctions. However, it is difficult to find research which attributes specific causes to the unsuccessful performance of these decisions and tasks.

Maintenance documentation and instruction are, for the most part, written by system designers at manufacturers like Boeing or Airbus, to be comprehended by maintenance personnel all over the world. This includes maintenance manuals, parts catalogues, service bulletins, structural repair manuals, troubleshooting manuals, final assembly instructions, and delivery instructions. These documents may be procedural, providing instructions for the correct execution of a task; informative, providing training materials; or legal, providing the obligatory proof that a task has been performed in accordance with regulations (Knezevic, 2015). Despite the operational need to use English to comprehend and interpret this material, training of maintenance technicians is often done in the national or local language (Drury, Ma, & Marin, 2005).

Fleets of aircraft from all airlines in the world are subject to potential mechanical errors caused by inadequate English proficiency resulting in the incorrect execution of a maintenance task. A report by the Transport Workers Union found that about half of operators of US-registered aircraft outsource their maintenance. Although the FAA requires certified mechanics to "read, write, speak and understand English,"[2] the FAA regulations do allow individuals to become FAA-certified if they are employed outside the United States by a US air carrier. In this situation, the certificate is stamped "valid only outside the U.S." (FAA, 2013).[3] However, those working at repair stations as supervisors must meet English language standards (Transport Workers Union, 2018). These low-cost maintenance facilities are in low labor cost countries with emerging economies like El Salvador, China, and Poland. Compounding existing and potential challenges related to adequate maintenance support, the projected world demand for 754,000 new technicians in the next twenty years, with the largest number again coming from the Asia-Pacific region at 257,000 (Boeing, 2018), should be cause for increased concern and efforts to ensure adequately trained and competent personnel to meet this growing need.

As a part of the solution to these challenges, Knezevic (2015) suggests that maintenance documentation be written in Simplified Technical English (STE), as noted in Chapter 5, to provide a more uniform language for the NNES around the world who must utilize these materials. ASD-STE100 recommends that all documentation be written in STE, but this recommendation is not typically or consistently followed. For example, while ASD-STE100 recommends the use of infinitives, imperatives, or simple present, past, or future verb uses, and explicitly states "Use only the active voice in procedural writing ... and as much as possible in descriptive writing" (STEMG, 2017), Udell (2018) found that the ten most frequent verbs in a Boeing 737–200 manual use passive voice. However, the writing or re-writing of this documentation would be perhaps just as time-consuming as quality translations, and would require an expert not only in the subject matter but also in STE. Knezevic cautions, however, that if the language of maintenance documentation continues to be uncontrolled, the risk of faulty execution of tasks will also continue. One can certainly argue that the language of pilot-controller communication has been controlled through phraseology, so it is also appropriate to implement a method of control and standardization in maintenance communication as well. Although the communication registers are quite different (text-based rather than voice-only), ICAO has prescribed strict guidelines as to what can and should not be said in RTF. Establishing equivalent ICAO standards for maintenance documentation and related training for maintenance communication would support safety in maintenance as the industry continues to expand in parts of the world in which English is not a national language.

7.8.2. Language training

Guidance in ascertaining the proficiency of inspectors, supervisors, and individuals authorized to issue approvals (the three positions in MROs *required* to "understand, read, and write" English) is provided by the Aeronautical Repair Station Association (ARSA, 2017):

- Ask the individual to read from English-written aviation materials and have her/him write an explanation of the materials in English.
- Provide the individual with verbal instructions in normal-paced English and have her/him write down the instructions provided in English.

The method of evaluating the resultant production is left up to the company or organization. Additional suggestions include asking individuals to fill out applications in person and to participate in conversational interviews. Considering the importance of matching test tasks with real-world, job-specific language use tasks, the validity of these assessment methods may be questionable. However, language instructors might use this information to design courses which would enable learners to successfully accomplish these tasks and meet the requirements to obtain employment, but it is essential to note that additional training in reading, speaking, and listening skills would be necessary for them to become strong enough communicators in the aviation maintenance environment to perform their job tasks accurately and competently at the level necessary to ensure safety in the industry. As stated previously, even with a poor testing instrument, quality training with MRO domains and tasks can have a huge positive impact.

More research is certainly needed to understand this register of aviation English communication. Corpus linguistics and written discourse analysis research as explored in Chapter 5 may be able to provide trainers with more insight into system designers' communication in maintenance documents which are used by technicians. Trainers who have a better understanding of these materials and the intended outcomes as they must be perceived and implemented by maintenance staff would be better equipped to facilitate accurate understanding and implementation by maintenance personnel during training and be able to better design and utilize training materials in their teaching to that end (Udell, 2018). Curriculum designers need more understanding of the lexico-grammatical choices made by technical writers to develop the curricula specific to maintenance that is still needed on the market. As of today, there are only a few English for Maintenance materials commercially available, but this specialized context is gaining more attention and traction, particularly with the increased focus of mainstream media on the outsourcing phenomenon and the personnel shortage reports issued yearly by Boeing.

7.9. English for flight training

7.9.1. Context

Globalization is causing more multilingual operational settings to appear where pilots, cabin crew, maintenance technicians, ground personnel, and ATCs are communicating using English as a *Lingua Franca*. Airlines are hiring crew members from a variety of countries who must work together, communicating in the common language of English, to provide safe and efficient flight operations for millions of passengers worldwide. For example, Emirates (2015) report having over 20,000 cabin crew staff composed of over 135 nationalities and speaking over 55 languages. These crew members must not only communicate with passengers, but also with other cabin crew members and pilots.

Pilot shortage. The 2018 annual Pilot and Technician Outlook published by Boeing reported that over the next twenty years, the world will need 790,000 new pilots. The greatest need is in the Asia-Pacific region where 261,000 will be required. Professor and Department Chair of Aeronautical Science at Embry-Riddle Aeronautical Science, Dr. Michael E. Wiggins, reported at a recent conference, "Right now the airline industry is facing as much stress in the hiring domain as it has in years. There is a big push to try to get flight students through the training faster, quicker, cheaper and more effective[ly] so that we can continue to safely operate the National Air Space System" (Circelli, 2018). As aviation continues to grow in these regions where it has not always been so prevalent, new training facilities and initiatives may experience this pressure to produce pilots quickly. As developing adequate English language proficiency inevitably requires significant investment of time and resources, there is a potential risk that pilots are accepted into training or even licensed without yet having obtained ICAO Operational Level 4 English proficiency. This is a risk, considering the unregulated nature of aviation English language testing.

7.9.2. Flight training in English

Many countries in regions like the Asia-Pacific do not currently have the training capacity for these 261,000 new pilots, and, therefore, send new cadets to countries such as the United States for education and training. In addition, high costs, restricted airspaces, and even the allure of what many perceive as a prestigious FAA license draw students to the United States for flight training. Many of the students coming to the United States for training are contract students, meaning they work for a company, such as an airline, that is paying for them to get their flight ratings and then return to their home countries to fly (Dusenbury & Bjerke, 2013). This puts pressure on training programs to be both fast and effective. As these flight schools continue to see increasing enrollment from international students, English has quickly become the default language not only of flight operations, but also of **ab-initio** flight training (Emery, 2015). [Discussed in Chapter 9, *ab-initio* flight training encompasses the initial training needed to acquire any pilot license, from the private pilot license (PPL), which is common for those interested in flying as a hobby and not to make money, to the commercial pilot license (CPL), which is the necessary license in order to have a career as a pilot.] Students are quickly immersed in English Medium Instruction (EMI) flight training programs, both on the ground and in the flight deck, expected to interact daily with their flight and ground instructors, flight school administration, and fellow cadets in a high-stakes, high-stress environment with many potential safety implications all presented or facilitated in English.

Admission criteria. Flight schools may admit NNES students based upon a variety of criteria. For example, collegiate flight training programs must comply with the admission criteria of their respective universities (Dusenbury & Bjerke, 2013), which include, most likely, a specified TOEFL or IELTS score. There are problems inherent in accepting such scores as indicative of readiness for flight training, since TOEFL and IELTS scores are interpreted by admissions departments as a single score. Individual sub-scores are not considered in the decision, creating the very real possibility that a student might score exceptionally high in skill areas such as reading and writing, but low in areas such as listening and speaking and, therefore, lack the sufficient oral communicative skills needed for flight training (Albritton, 2007).

While the FAA (2013) does provide limited guidance in AC 60–28 as to how an examiner should evaluate a student pilot's English ability during training, it does not provide guidance as to how these flight schools should screen applicants initially. Flight schools may base their admission decision on tests such as TOEFL or IELTS, or they may require an in-house admission test or interview, but with no regulations or specifications as to the design, administration, or content of such assessments. Regardless of what may or may not be the case relative to the various admissions criteria or protocols themselves, there is no research-based benchmark yet identified for entrance into flight training, so regardless of the testing instrument's validity, the acceptable score for these candidates has still not been determined. However, oral exams taken prior to flight training have proven to be significant predictors of success. Dusenbury and Bjerke (2013) found that Chinese flight students who completed flight training had significantly higher scores on a pre-flight training oral exam than those who did not complete flight training. In addition, those with higher scores completed their private pilot training with a lower number of total hours required.

Placement tests. ICAO (2009) explicitly states that placement tests should be used to position students at the right stage in their training program (referring to professional pilots and controllers). These types of test instruments should contain reference only to subject matter that specific populations of students are familiar with (meaning *ab-initio* and professional students require different tests). However, the role of ICAO itself is to develop and maintain Standards; it is not the role of ICAO to create assessment instruments. Language testing is still a widely unregulated industry within the aviation world. It can be argued that, minimally, there needs to be a context-specific assessment instrument for each of the positions identified above, with both ab-initio and professional versions available. More development in language testing for aviation is certainly needed.

An ab-initio pilot test. A general test assessing speaking and listening skills, such as an Oral Proficiency Interview guided by the American Council of Teaching Foreign Languages (ACTFL), can be given prior to flight training to determine readiness. Semi-direct methods have been found to be effective and cost- and time-efficient, but these must be conducted by trained and calibrated raters familiar with ICAO rating scale. It is of utmost importance, however, that airlines and flight training organizations standardize their methods of pre-flight training testing (Albritton, 2007), in contrast to the ad hoc tests which are being utilized currently. A pre-flight *ab-initio* test should not presume aviation knowledge or the ability to use phraseology in radiotelephony communications, particularly if this test is to be used to qualify for entrance into a training scenario in which students do not yet have exposure to this knowledge (Emery, 2015). The exams used in the professional pilot-controller domain, as ratable by ICAO LPRs, are therefore unsuitable as determinants for entrance, as quality ICAO language tests will test pilots and ATCs in a context similar to the one in which they work, with content relevant to their work roles (ICAO, 2010, Section 6.2.8.3).

The six required ICAO language components can still serve as a basis for *ab-initio* pilot language evaluation, but adapted to the particular context. Some institutions couple this type of assessment with a written general English vocabulary and grammar test to assess a learner's general English level competence absent any aviation bias (Bullock, 2015). Based on the results of the pre-flight *ab-initio* pilot test, various plans of action can be employed. Some institutions may wish to only accept candidates who are already at the necessary level of language proficiency, but the reality is that the disparity within formalized recruitment and admission criteria mean that many *ab-initio* trainees will be admitted and then need supplementary language training before their professional courses begin (Mell, 2004). Flight schools cannot afford to turn away such a large number of students. Albritton (2007) equates ICAO rating scale to training paths as illustrated below in Table 7.2. (Note ICAO

Table 7.2 ICAO rating scare and training/language path (adapted from Albritton, 2007)

ICAO Rating	Language Path
Operational Level 4 or higher	Would be able to begin flight training, but could benefit from specialized language support during their flight training, such as intensive ATC communication practice and possible individual language tutoring on an as needed basis.
Strong Pre-Operational Level 3	Would be able to begin flight training within an estimated 4–8 weeks of intensive 5–6 hour per day/5 days per week of communicative aviation English language training. The course should be content-based using authentic materials from flight training in order to complement ground school. Student progress will vary depending on the language learning environment, the student's linguistic background, education and personality factors.
Weak Pre-Operational Level 3	Would be able to begin flight training within an estimated 12–16 weeks of intensive 5–6 hours per day/5 days per week of communicative aviation English language training. The same content-based materials could be used as for strong ICAO Preoperational Level 3; however, more specific ESL modifications would need to be made and additional activities added to the curriculum in order to deal more effectively with the students' lower proficiency. Student progress will vary depending on the language learning environment, the student's linguistic background, education and personality factors.
Elementary Level 2	Would likely require a minimum of 6 months or more of content-based aviation English and general English in an intensive, highly communicative language environment. Student progress will vary depending on the language learning environment, the student's linguistic background, education and personality factors.

rating needs to be assigned based on an adapted scale and testing instrument appropriate to the context.)

Language needs for flight training. Participating in and developing the requisite skill level in the restricted register of RTF, replete with fast-paced exchanges and phraseology, is a major challenge for NES, and can be even more challenging for NNES, particularly if their language skills are inadequate. Furthermore, *ab-initio* pilots need to prepare for a variety of language tasks, including situations such as debriefing with a flight instructor, reporting technical problems, speaking with a dispatcher, and receiving non-standard clearance from ATCs (Albritton, 2007). Flight training also requires receptive reading skills for ground school, checklists, handbooks, manuals, and other situations and applications, and many of the same skills needed in academic English, such as focused listening and note-taking.

Once training begins, instructors, administrators, and the students themselves may be surprised by the apparent lack of English capability many students face to meet these high communicative demands of flight training. In many such situations, the task of fixing these language issues falls to the flight instructor, who has no formal training in language acquisition to diagnose, manage, or address the problem of insufficient language proficiency. This might result in delays in completion of training and higher-than-expected expenses for the trainee. In addition, this situation clogs the flight training system with the short supply of instructors already strapped for time. The FAA Aviation English Language Standards (AELS) described in AC 60-28b stipulate that students in in-flight training can broadly "read, speak, write, and understand the English language." AC 60-28b requires that flight schools, CFIs, and check airmen "continuously monitor" flight training candidate compliance with FAA AELS Operational English language proficiency level. The FAA emphasizes the responsibility that training facilities and flight instructors share to ensure that airmen meet the FAA AELS. However, this regulation is often interpreted loosely, as even if flight schools do recognize their lack of compliance with FAA requirements and their recurring language issues, there are few scalable solutions widely available which have been proven to produce reliable results.

7.10. The aviation English instructor

Although online aviation language learning is making strides toward potentially effective asynchronous, self-paced training solutions, most aviation language training programs are still done with the aid of an aviation

English instructor or teacher. Each of the above contexts needs dedicated instructors for both initial and recurrent language training. The field of aviation English teaching is unlike other language teaching fields in many ways, and therefore requires specialized training to ensure requisite competencies of an instructor. Petrashchuk (2015) outlines three important distinctions. First, aviation English teaching is a safety issue, the results of which—favorable or unfavorable—will quite likely impact the livelihood of trainees and, in the case of aviation, the passengers aboard as well. It is critical that instructors understand the potential that language may have on already-stressful and safety-critical situations, particularly in pilot-controller voice-only communication, and the possibility of fatal consequences from a seemingly small misunderstanding or mistake (ICAO, 2009). In addition, aviation English contains both the restricted register of phraseology and the broader, "natural" register of plain English in an aviation context. It is important to note that even this "plain English" is still highly specialized, situation-dependent, and simplified according to the same requisite conciseness as aviation language. The teaching of such language requires a certain degree of familiarity with the language itself, particularly when facilitating communicative activities. Finally, ICAO language requirements and the tests which accompany them are certainly a part of the total reality for many aviation English students, and instructors should be cognizant of the impact poor performance as an instructor can have on students' ability to pass such tests and qualify, therefore, to work as professionals in the aviation industry. The high stakes related to ICAO language tests can also cause anxiety and/or powerful motivation for "test prep" by the students. In general, the stakes in aviation language teaching and learning are higher than many, if not most, other ESL/EFL/ESP settings.

7.10.1. The roles of an aviation English teacher

Along with other ESP professionals, the aviation English teacher often must accommodate and be productive within several distinct roles within the larger infrastructure of aviation English. (Refer to the Pyramid of Infrastructure earlier in this chapter.) Of course, participation in the development and implementation of training programs is typically the primary focus of work for such a professional. However, in the case of aviation training, responsibilities will often extend beyond the student pilot or maintenance technician to the training of other teachers also. Within a single flight training institution, for example, one person may be responsible for the creation of a placement exam, the administration and scoring of this exam, the development of training courses, and the teaching of these courses. In addition, administrative tasks are also a common part of the responsibilities due to the programmatic nature of aviation English courses which includes not only the training component, but also a testing component as well. Students need to be tracked, monitored, and re-tested over the course of the training program. Aviation ESP professionals also work in consulting, advocacy, and policy analysis and development, all crucial components of successfully building and maintaining the necessary training infrastructure. (Refer again to the Figure 7.1 illustration of the recommended infrastructure.)

7.10.2. Instructor qualifications

It is not uncommon for English language instructors to find themselves in aviation teaching roles without the necessary background knowledge, in which case they must become familiar with the essential industry-specific knowledge, including operational language, necessary to fulfill their training responsibilities. English teachers are not typically well-versed in industry operations and must, therefore, be trained to apply the knowledge they have gained from their academic background in a new and very industry-specific context. The amount of operational knowledge that an instructor should have has not been systematically investigated and defined, but instructors should minimally have foundational knowledge, such as aircraft systems and equipment, flight operations procedures, components of an airport, aviation weather, and radio communication. Instructors with only minimal knowledge in these areas are at risk, potentially a safety risk, of misguiding students in the classroom and/or compromising their credibility in their instructor role with students. For this reason, instructors must be confident enough to recognize their *lack* of aviation content knowledge and, if necessary, to admit to not knowing and seek support from other professionals or resources. It is always a wiser decision in such a high-stakes safety environment for the instructor to divert the question to the next class to allow time for the necessary research to provide an accurate answer. Many training and/or limited certification programs do advise that aviation English instructors have access to an operational subject matter expert such as a Certified Flight Instructor (CFI) or an airline captain when the need for consultation arises in the course of instructing students.

ICAO Document 9835 (2010) provides qualification guidelines for aviation English instructors, test developers, subject matter experts, and other team members, summarized in Table 7.3.

In the context of the global aviation industry that ICAO represents, it was important to provide a path to aviation English teaching that included "minimum qualifications," as there are areas in the world which would not have access

Table 7.3 ICAO guidelines for aviation language trainers, administrators, and materials developers (adapted from ICAO Document 9835, 2010)

	Qualifications		
	Best	Very Good	Minimum
	1. Aviation language trainer, administrator, or materials developer		
Language training academic qualifications	Master's in Language Teaching • Teaching English as a Second Language (TESL, TESOL) • Applied Linguistics • Foreign Language Education or a related field	• Bachelor's degree in foreign language training, or • Graduate diploma in TESL, etc., or • University degree + Extensive L2 or foreign-language training experience with clear evidence of commitment in the field	• Certificate in TESL, or • University degree (initial training should be done under close supervision of experienced trainer)
Language training experience	Aviation language programme 3+ years	• Aviation language programme • Language for specific purpose training • Language training in an accredited university of language school	• Language training experience, or • No previous training experience acceptable when training is under close supervision of experienced trainer
Aviation communications	Pilot or controller experience	Radiotelephony familiarity (through aviation language apprenticeship or experience)	Ability to work well with SME
Language learning materials development	Aviation language materials development with communicative or interactive approach		Language learning materials development with communicative or interactive approach
Language training administrative experience	Aviation language program administration	Language training program administration	Aviation or language training program involvement

to language teachers with Best Qualifications. However, in English-as-a-first-language contexts and for *all* commercial providers of aviation English training, teacher qualifications should represent "Best Qualifications." ICAO notes that other "language-related" academic fields such as literature, communications, or even "pure" linguistics do not focus enough on language learning or training and are, therefore, not relevant qualifications for aviation language. In order to obtain the radiotelephony familiarity described in the table, ICAO recommends that language trainers utilize a variety of means, including "taking flight lessons or observer flights; through an apprenticeship with an aviation language master training; through experience in teaching aviation language; through interactions with professional aviators and controllers; through reading widely and other self-educational schemes; and through the use of simulators and software programmes." ICAO (2009) provides the following list of activities that aviation English teachers, or those intending to become so, can take part in to prepare themselves for the challenge of teaching aviation English:

- Listening practice in an ATC lab
- Utilizing ICAO Rated Speech Samples Training Aid
- Developing and delivering communicative language lessons from raw data
- Adapting lesson plans and content to meet specific needs
- Observing and working in tandem with experienced trainers

These recommendations are useful for any aviation English teacher, but, in particular, instructors working with pilots or controllers. Instructors who work with maintenance technicians will need another knowledge set more specific to maintenance documentation. The acquisition of essential knowledge in this domain is indeed challenging, due, among other things, to the fact that more research is needed to understand exactly what comprises the register of essential maintenance documentation in order to then develop appropriate language objectives, teaching materials, and strategies for training maintenance technicians.

Even after the initial acquisition of essential working knowledge about the aviation industry, the function of the components comprising it, the roles and responsibilities of each category of employees and the skills related to the different tasks within the domain, teachers still need specialized, recurrent training. Although not always recognized and/or provided in the aviation world, language trainers need a sustainable program set up to facilitate and support on-the-job, ongoing training, including frequent contact with operations experts, access to aviation training materials, and dedicated time to maintain familiarity with new developments, policies, or training procedures (ICAO, 2009).

As mentioned above, the acquisition of the *best* qualifications as outlined in the table is typically not achieved through MA programs in applied linguistics or TESOL. A more suitable and productive path is to first obtain an MA in applied linguistics or TESOL and then proceed to gain actual working knowledge of the industry in general and its various components in particular through one or more of the strategies suggested above. ICAO Circular 323 (ICAO, 2009) suggests that a minimum of three years' experience is necessary before a trainer has enough familiarity to be competent and comfortable enough with the aviation subject matter and domain to confidently and effectively train employees in the various roles and responsibilities. Currently, new initiatives are being undertaken to help better prepare English language teachers to become specialized aviation English language teachers, such as training programs and internships at aeronautical universities during TESOL or applied linguistics post-graduate degree programs.

Working with an operational expert. At times, it may be necessary for an English language trainer to work in tandem either with another, more experienced language trainer or with an operational expert (1) if the trainer's aviation knowledge is not sufficient, or (2) if it is part of the trainer's own training requirements, or (3) if the instructor's knowledge of language training techniques requires the presence of a more-qualified instructor (ICAO, 2010). The presence of an operational subject-matter expert (SME) in the classroom has its own advantages which language trainers may not fully appreciate until they encounter complex, challenging technical inquiries from students. A course which has successfully integrated roles for both a language SME and an operational SME is English for Visual Flight Rules (EVFR). The resultant benefits of this model are outlined in Chapter 9. If a program is designed to have dual SMEs, it is helpful to first cross-train the team by immersing the language SME in training to provide an overview of the aviation domain while, at the same time, providing the aviation SME with knowledge and experience in the language teaching domain.

Teaching and testing aviation English are both highly specialized functions which require considerable professional and academic expertise. Aviation English is not the right field for those who are unable or unwilling to first commit to obtaining basic qualifications and levels of proficiency and to continue professional development in the field, including staying abreast of constantly emerging developments in both language teaching and the aviation industry. Instructors will be expected to demonstrate, in all aspects of their instructional and organizational performance, a comprehensive understanding of their key role in supporting aviation safety and a commitment to the high professional standards demanded of aviation professionals employed in the industry in the interests of service to and safety of the passengers.

7.11. Conclusion

Much work remains to fill the infrastructure gaps surrounding full compliance with the ICAO LPRs as directed to professional pilots and ATCs, as well as the English language needs of other aviation personnel outlined above. When considering all the necessary components needed, from tests with trained and calibrated raters, to training programs staffed by qualified language instructors, the great amount of work remaining is clear. As aviation continues to grow, training and testing will also continue to gain traction, and with this growth comes a great opportunity for TESOL and applied linguistics professionals to develop innovative and effective solutions and to conduct seminal research into the domain of aeronautical communication. Questions such as the language proficiency needed for flight training and the register specificities of maintenance documentation remain unanswered. Systematic, thorough, and carefully designed aviation English training programs are needed with full financial and logistical support from the organizations providing such training. Aviation employs almost 10 million individuals, many of them second language users of English, and as this industry continues to grow, the need for aviation English training grows as well.

8

The Development of Aviation English Programs

with Malila Prado and Patricia Tosqui-Lucks

8.1. Introduction: Pedagogy in aviation English

The development of aviation English programs requires careful consideration of pedagogical principles, together with a focused and extensive needs analysis. Program administrators make choices regarding language objectives, content sources, curricula, materials/activities, detailed modalities, and duration of learning—choices which should be based on the specific operational context of instruction. To illustrate, this chapter provides an overview of relevant pedagogy in the field of aviation English teaching and learning, situated in the areas of English for Specific/Occupational/Academic Purposes (ESP/EOP/EAP) and teaching approaches such as communicative language teaching (CLT), content-based language teaching (CBLT), and task-based language teaching (TBLT). This chapter also outlines putting theory into practice in the industry, including sample lessons of customized aviation English courses.

8.2. English for Specific Purposes

Aviation English is unquestionably considered part of the broader field of English for Specific Purposes (ESP). ESP is defined as "the area of inquiry and practice in the development of language programs for people who need a language to meet a predictable range of communicative needs" (Swales, 1992, p. 300). ESP includes a range of industries from medicine to business to law. Arguably, these ESP industry areas have traditionally been the contexts of more research studies, materials development, and even teacher-training than has aviation; therefore, there are today many still-unfilled gaps in the infrastructure of aviation English. Despite the obvious and even urgent need for aviation personnel to be able to communicate accurately to ensure the safe and efficient transport of passengers on aircraft around the world, and even with the development and subsequent implementation of ICAO standards, the engagement with aviation ESP within the academic TESOL community remains limited. The construct of aviation communication still needs closer, research-based examination in order to inform the most effective and efficient pedagogies for both training and testing. While the specific purpose of aviation communication is clear, the English language component and related issues are still extensively debated, focusing on such issues as the composition of this specialized language, its users and the regulations imposed on its users, the best methods for teaching and assessing, and even the final decision on nomenclature (e.g., aviation English, aeronautical communication, aviation language, or *airspeak*).

8.2.1. ESP and aviation

The language of aviation certainly fits Swales' definition of ESP above, particularly in reference to radiotelephony communications. Communications in the flight deck have a "predictable range," as prescribed by the phraseology found in ICAO's *Document 4444: Procedures for Air Navigation Services—Air Traffic Management* and *Document 9432: Manual of Radiotelephony* (2007) and the aviation plain language used in non-routine situations. However, if RTF is considered the only domain of aviation English, then all students of flight and air traffic control, and *only* those of flight and air traffic control, learn this language. Since even native speakers of English must learn this "language," aviation English fits the definition of a *lingua franca* in that it is used by speakers without a common language, supporting the argument that aviation English has no native speakers (Estival, Farris, & Molesworth, 2016). But, is the language which is learned by NES to perform their job appropriately termed "aviation English?" It is certainly true that native English speakers can benefit from instruction in effective communication in an aviation environment, not only in the specifics of RTF phraseology and technical terminology, but also including aspects of strategic competence, such as an awareness of the potential misunderstandings in intercultural communication and sensitivity to strategies to confirm comprehension (ICAO, 2010). Little research has been done on what NES should be taught (Estival, Farris, & Molesworth, 2016).

However, although NES and NNES ultimately are expected to produce the same universally understood language in aviation contexts, for this discussion, the focus is on NNES and the teaching and learning of language and communicative competencies of all aviation personnel whose first language is not English, the international language of aviation.

Aviation English includes such a variety of contexts and communication scenarios that it is best to first think of it holistically in the same way as any other subdivision of ESP, such as English for Medicine or English for Business. Unlike business English, though, aviation English is unlikely comprehensible to even a NES without proper training, as mentioned above. Whereas an outsider or layperson hearing business English could likely understand the utterances if not the complete meaning, an outsider hearing or reading a pilot-controller transmission will most likely not understand most of what was said (Estival, Farris, & Molesworth, 2016). These assumptions of comprehensibility or lack thereof are particularly true in RTF scenarios. A non-specialist listener might be able to understand a conversation between a maintenance technician and a pilot, albeit perhaps not understanding all the technical vocabulary, more clearly than a transmission between a pilot and a controller on the radio. RTF as its own domain is quite unique, as it is a highly restricted register and used only in a specific workplace communication setting, that is, the aviation environment.

As stated above, within the domain of RTF, there is another subset of language upon which ICAO LPRs (ICAO, 2010) focus in particular: radio communications between pilots and ATCs occurring when non-routine events occur. The term applied to this specific language which is used in unusual circumstances or events in LPRs is "plain English" or "plain language." (See Bieswanger, 2016; Lopez, 2013; Moder, 2013 for further discussion.) For the routine phases of the flight, RTF must be used as documented (ICAO, 2001; ICAO, 2007). Nevertheless, if and when unexpected situations or events occur, plain language should be used in the same format as RTF, that is, "clear, concise and unambiguous" (ICAO, 2010, p. 4.2). Plain language is required not only for urgent, unusual, or emergency situations, but also for routine communications when pilots and controllers need to share information, clarify or confirm an issue (Doc. 9835, 2010, 3.3.17).

8.2.2. English for Academic Purposes or English for Occupational Purposes

English language training can be divided into the categories of General English and ESP. ESP is then sub-divided into English for Academic Purposes (EAP), including fields such as business, engineering, and law, and English for Occupational Purposes (EOP) for industry sectors, NGOs, and government (Lomperis, 2018).

The placement of aviation English is therefore tricky, depending on the context of aviation English being considered. (See Chapter 7 for a description of aviation English contexts.) For example, flight students who are studying English either before or during their flight training programs need more of a traditional EAP-focused curriculum, preparing for their present and future training needs which will necessarily involve listening comprehension, reading strategies, oral presentation abilities, and even study skills to be successful in the "classroom" of flight training. (See Chapter 9 for a description of three English for Flight Training courses.) In fact, the *ab-initio* language training programs for pilots, air traffic controllers, or maintenance technicians may all begin as more Pre-Academic General English (Lomperis, 2018) to obtain the necessary academic language skills needed in a training environment. As learners progress through training and subsequently enter the workforce, the language training will also evolve. For example, maintenance technicians who are already employed and taking part in recurring English language training need more of an EOP curriculum which includes a focus on tasks such as record-keeping or the comprehension and application of information in manuals. In addition, training targeted at RTF is also categorized as EOP since its components do not, in isolation, represent a "language" which can be utilized in contexts outside of the workplace environment (Aiguo, 2008). If a course combines the teaching of RTF and aviation plain language to prepare for a training environment, it is then considered to have elements of both EAP and EOP.

Regardless of the EAP/EOP distinction, it is important to note that aviation communications are fundamentally strictly defined, requiring standard, precise, and accurate use of phraseology, for example, in RTF, aircraft maintenance terminology and specifications, and even legal terms in aviation law (Aiguo, 2008). However, successful communication in aviation contexts also requires the general use of the English language in daily communications not only once professional employment has been attained but also even before employment in training scenarios at flight institutions, universities, and other training facilities.

8.2.3. Target language use tasks

According to Hutchinson and Waters (1987), ESP should be learner-centered and needs-based, with collaboration between workplace and content experts, and utilize authentic tasks and materials. When designing an aviation English program, these defining characteristics must be taken into consideration. ICAO (2009) states that a curriculum must aim to ensure that learners are acquiring the necessary skills to achieve operational proficiency in their real-world professional lives and must simulate as closely as possible the operational communicative functions used in their jobs.

Although ICAO LPRs do not refer to *ab-initio*, maintenance, cabin crew, or ground staff trainees, curriculum designers should use these same guiding principles and consider the real-world communicative functions of their environment as well. In an English language training program for *ab-initio* aviation professionals, the types of tasks and language skills should be like those typically found in any other *ab-initio* training environment, including a focus on such objectives as understanding class lectures, taking notes, answering questions from instructors, participating in discussions, and giving oral presentations. These types of tasks mirror the types of target language use present in operational training, for example listening to and taking notes on a lecture about oil systems, answering questions about real-time decisions from a Certified Flight Instructor (CFI) during a flight, or participating in oral examinations during final check rides.

On the other hand, the daily tasks of professional aircraft mechanics, for example, are similar to those commonly found in EOP contexts, such as writing reports or reading manuals or for maintenance or repair of equipment. After becoming a professional, employed technician, the focus will be more on workplace performance. For example, in the case of a commercial airline pilot or a professional air traffic controller, curriculum designers must focus on specific work-related communicative functions, for example, those related to RTF, such as the constraints of voice-only communication, the immediacy of responses, dealing with unexpected situations or sudden changes, resolving misunderstandings, etc. (Mell, 2004).

An additional useful distinction from the work of Knight, Lomperis, van Naerssen, and Westerfield (2010) is the division of need for English language in ESP. Students in areas such as flight training fall into "Category 1," or learners in the process of developing expertise in their fields who need to be able to use English as a tool in their training, while commercial airline pilots fall into "Category 2," learners who already possess a great deal of expertise in their fields and need to use English as a tool in their work. When designing curriculum for these two sub-groups, special attention to the content material and the target language use tasks must be given in order to ensure the material is relevant and expertise-level appropriate for the students, as their target language use needs are related, but different enough to warrant tailor-made materials for specific contexts and ability levels.

8.3. Communicative Language Teaching

Communicative Language Teaching (CLT) can be described as an approach focused on interaction, fluency, and on using language for a meaningful purpose (Brown, 2007). Lesson particulars generally follow a sequence of "Present, Practice, and Produce." During the Present stage, students are introduced to the material with explanations of content. The teacher provides the target language needed to accomplish a goal. In the Practice stage, students are given more control and opportunity to use the language, but in a controlled and structured manner. Finally, in the Produce stage, students use the language freely, but follow guidelines previously established. Because aviation English is rigid, strict, and full of standardized discourse, some aviation English instructors tend to easily fall into a more drill-based and form-focused methodological approach in the classroom. However, the CLT approach is highly recommended here as students will, after all, be using this language to communicate, primarily verbally. Facilitating activities which do not demand communication may be useful for brief instances when students are being introduced to content, but the ultimate goal in an aviation English classroom should always be to have the students actively working with the content to produce meaningful interactions.

An additional conceptual framework for CLT, focusing on speaking, is Thornbury's (2008) three stages of learning: awareness-raising (present), appropriation (practice), and autonomy (produce). Each of these stages makes use of specific activity types. Activities in awareness-raising include the use of recordings, transcripts, and live listening, all of which should be authentic or derived from actual sources. Activities in the appropriation portion include dialogues, jigsaw activities, gap-fill activities, and assisted performance. Finally, in the autonomy stage, learners participate in presentations, role-plays, discussions, and debates.

Aviation English can be categorized into four speaking events: *transactional* (exchange of services), *interactive* (pilot/controller), *planned* (normal flight), and *unplanned* (unexpected events) (Bullock, 2015). These communicative functions can guide instructors when choosing techniques to use in the classroom. In teaching aviation English, real-life situations, communicative competence rather than reliance on form, pragmatic contextual content, and the ability to accomplish appropriate communication are seen as the most important areas to cover (Bullock, 2015) through the communicative approach.

8.4. Task-Based Language Teaching

Task-Based Language Teaching (TBLT) utilizes tasks to accomplish meaningful learning objectives (Bygate, Skehan, & Swain, 2001). Skehan (1998) defines a "task" by four criteria: (1) meaning is primary, (2) a goal is set to be accomplished, (3) there is an outcome to be evaluated, and (4) a real-world relationship exists between the classroom task and outside goal. Tasks can be divided into *target tasks*, which are those tasks students will be asked to do outside the classroom, and *pedagogic tasks*, which are those tasks which students will be asked to do in the classroom to prepare for the target tasks. Ultimately, pedagogic tasks should aim to simulate target tasks through activities such as role-plays. A curriculum which follows a TBLT framework will identify target tasks and create pedagogic tasks which seek to help students accomplish the final goal outside of the classroom (Brown, 2007). In a task-based approach, successive attempts are made at accomplishing a task, with input from the instructor given at each step which draws attention to the linguistic components of a task and allows for improvements. As the ultimate goal is for learners to develop the capacity to utilize these language features in real-world scenarios, the class can be set up to provide learners with awareness and then opportunities for integration of the language into the task (Paramasivam, 2013).

TBLT is particularly effective in aviation English classrooms to introduce, as much as possible, the real-world target language use scenarios in an artificial environment before these tasks are required in operation. Through simulating target tasks, pilots and air traffic controllers can prepare and build confidence with the freedom to pause, ask questions, confirm, and start over. TBLT is akin to scenario-based training which is increasingly becoming the preferred method of training in operational settings as it provides opportunities for personnel to practice their aeronautical decision-making skills. Simply retelling a scenario with its results is not an effective method and does not necessitate communication. Instead, scenarios can be given which require learners to negotiate, delegate, discuss, and finally complete pedagogic tasks which are, as closely as possible, accurate representations of operational target tasks.

8.5. Content-Based Language Training

ICAO Document 9835 (2010) outlines in Section 7.5 that Content-Based Language Training (CBLT) is the recommended approach when training pilots and air traffic controllers. The following rationale is provided:

a) Many aviation safety organizations publish and distribute videos and tool kits designed to improve the safety awareness of pilots and air traffic controllers. Much of these materials are freely distributed and publicly available. Most are published in English, and target English speakers at what ICAO calls the "Expert Level 6".

b) Many pilots and controllers with limited English are not able to easily access the safety information contained in such publications.

Adapting such publications for an aviation English program makes the information contained there accessible to all pilots and controllers.

c) Improving safety awareness is an ongoing process. Pilots and controllers universally exhibit a high interest in increasing their safety awareness. In fact, organizations commit significant resources to the continual improvement and management of safety systems.

d) Materials are intrinsically interesting. Pilots and controllers who need to comply with ICAO Operational Level 4 may require between 200 to 400 hours of aviation language training. Providing content-based, safety-focused language training has a number of benefits for the pilots, controllers, their organizations, and the aviation industry:

   1) It doubles the value of required language learning time by pairing language lessons with important safety content.
   2) It increases safety awareness.
   3) It provides high-interest topics in the language lessons, increasing learner motivation.
   4) Motivation is a key factor in language-learning success. People naturally pay more attention to topics in which they have an inherent interest.
   5) Time spent on language learning has a positive impact on progress.

8.5.1. Approaches to using content

Language learners have traditionally acquired foreign languages because they needed to understand important content necessary for trade, business, settling into a new place, etc. (Mathews, 2014). In general, all forms of instruction which utilize some piece of content as a foundation combine, in some capacity, language learning with the acquisition of useful informational content. Considering the wide variety of contexts in aviation English, it may be helpful to briefly discuss the distinguishing features between English medium instruction (EMI), content and language integrated learning (CLIL), and content-based instruction (CBI). Brown and Bradford (2016) succinctly describe the differences by stating, "In EMI, content is central; CLIL has a dual focus on content and language; and in CBI, content is peripheral" (p. 332). A more detailed discussion of these three approaches and their potential applications in aviation English follows.

English medium instruction (EMI). Largely due to internationalization, EMI is increasingly prevalent across the world, primarily at the tertiary level. Descriptions of EMI do not typically make direct reference to the aim of improving students' English since the focus is on content-mastery. While there may be some language support, English is viewed as a tool for learning subject content and is not part of learning outcomes or assessment (Brown & Bradford, 2016). One of many possible contexts in aviation is students in flight training institutions where all subject matter is taught in English. In essence, this is an EMI course, since students will participate fully in flight training using the English language for the acquisition of knowledge and productive proof that the content has been mastered. Recurrent pilot training courses, or jet transition training, when pilots attend intensive ground school and simulator sessions to learn systems on a new aircraft, are also EMI courses. Ground and flight instructors, as well as simulator software, typically do not take time to scaffold or slow down the presentation of materials for NNES, a reminder of the importance of English proficiency prior to beginning a purely operational training scenario. There is no specific accommodation for English as a foreign language speakers. Additionally, the vast collections of important aviation safety materials, expected to be learned and put to use by aviation professionals, are typically published in English. For example, Flight Safety International offers a selection of online training modules, all in English, which are valuable resources for professionals to learn the latest industry information. Often, such online modules are required as part of recurrent training. The option to acquire the information in an alternative language is not possible; in this situation, English *must* be the medium of instruction.

Content and language integrated learning (CLIL). In CLIL, both the content and language are concurrently addressed through a focus on all four core skill areas (reading, writing, listening, and speaking). Grammatical sequencing is replaced by a more lexically focused exposure to language; there is a focus on form only as it relates to comprehension and production of the content. Additionally, CLIL uses content which is not merely a haphazard collection of current affairs or material of general interest; the content should, rather, be relevant to the curriculum of the larger program. Successful CLIL increases intrinsic motivation and avoids the artificial separation between language learning and real, subject-related communication (Mathews, 2014). Learning outcomes should include items related to both content and language, which necessarily means that students will be assessed in both areas as well (Brown & Bradford, 2016).

Aviation English courses, particularly those for professional pilots and air traffic controllers, can be designed quite effectively using a CLIL approach, particularly when the aviation English course is fully supported programmatically. The course can prepare for, support during, or reinforce after operational or safety content learning. For example, if a flight school made the conscious decision to teach ground school in a more language-focused, scaffolded way, it could be considered a true CLIL course wherein students learn English and learn the content of ground school simultaneously, in much the same way that many universities offer introductory English 101 writing courses specific for NNES.

In order for CLIL in aviation to be effective, the aviation English instructor must become familiar with the operational or safety content of the topic. Close collaboration with flight instructors in the development and delivery of CLIL in aviation ensures that the operational content is accurate. Finding an aviation English instructor who has a high degree of aviation operational familiarity is challenging. However, with the right degree of and time for preparation and, perhaps most importantly, the operational experience or access to resources to obtain it and support of an operational subject matter expert a language teacher can teach a CLIL aviation course quite effectively. Even if the course is not intended to be part of the larger curriculum, CLIL is still an effective approach in designing courses which will capitalize on intrinsic interest. Students are motivated by knowing that the course is introducing not only technical vocabulary but also operational concepts that they will be expected to use in real-world target language use scenarios. Although the larger program may not rely directly on the CLIL aviation English course as the exclusive teaching of operational content, courses can still be designed as if this were true with nothing to lose and a great deal to gain.

Content-based instruction (CBI). CBI is used often as an umbrella term to include any teaching of content in a second or foreign language. In this approach, the subject matter is used as a vehicle for learning language (Brown & Bradford, 2016), and the curriculum is centered on language learning outcomes and assessments. The content and language objectives should be cohesive, allowing units to flow seamlessly into one another to create a logical sequence in the course plan. CBI is categorized into three models identified by Stryker and Leaver (1997) (as cited in Brown & Bradford, 2016):

1 Sheltered content classes in which subject-matter learning is the primary intended outcome;
2 Adjunct-CBI in which language classes are taught parallel to subject-matter classes, or team-taught by content and language SMEs;
3 Theme-based CBI in which language classes are taught by language teachers using subject-matter themes.

CBI is present in many contexts in aviation English, particularly in *ab-initio* settings. CBI may also be utilized in aviation English courses which are part of university-based systems or specialized IEPs. One of the courses described in Chapter 9, Aviation Topics, utilizes a theme-based CBI approach to language learning. A course which follows the adjunct-CBI model is also described in

Chapter 9, English for VFR Flight, although this course is closer to CLIL in terms of the attention given to language and content.

Choosing the content. For aviation professionals to be successful, they need to be familiar with a specialized set of vocabulary and concepts which a non-aviation professional probably does not know. Shared knowledge of topics such as weather, aircraft components, airspace, and airports is what allows pilots, air traffic controllers, and aircraft mechanics to successfully communicate with one another (Mitsutomi & O'Brien, n.d.). Aviation English classes, therefore, should make use of this necessary shared content. Additionally, utilizing this particular content in the classroom works to satisfy the holistic descriptor which requires pilots and air traffic controllers to possess the ability to communicate on "common, concrete, and work-related topics" accurately and clearly (ICAO, 2010).

When choosing content for an aviation English class, it is of the utmost importance that curriculum developers be cognizant of both the operational knowledge level and the English skill level of the learners. Unless tandem-teaching, instructors are responsible to handle the operational content material with confidence and accuracy and must therefore choose it carefully (Miller, 2001). Additionally, developers need to choose content which is appropriate for their context and is compellingly interesting. Content which is not interesting and directly relevant to learners, even in "standard" English language programs, results in students feeling bored, frustrated, and unengaged. Content must also be relevant to the student population's actual profession (pilots, air traffic controllers, maintenance technicians, cabin crew, and so on) (Kovtun & Simoncini, 2014).

To illustrate, consider choosing content for pilots. Content which is considered foundational in the aviation industry, such as the parts of an aircraft or the four forces of flight, is appropriate in the following scenarios:

- *Ab-initio* English language training, regardless of ICAO level. Students may know this content in their first language, but the unit will still be effective when conducted in English.

- English language training for professionals who have Elementary ICAO Level 2 or low Pre-Operational ICAO Level 3. Students may be able to use this content as a launch pad to boost their confidence and bolster fluency, pronunciation, and vocabulary skills.

However, foundational aviation content is not appropriate in programs which contain professionals with a great deal of experience and high Pre-Operational ICAO Level 3 or higher proficiency. For them, this content in English will be too rudimentary and discourage engagement. Instead, they require more sophisticated, advanced topics such as High-Altitude Upset Recovery or Controlled Flight into Terrain.

Nation and Newton's (2009) framework for teaching listening and speaking is a useful model when considering syllabus design, materials development, and content selection for a specific course (as cited in Paramasivam, 2013). The framework consists of

1 Meaning-focused input
2 Meaning-focused output
3 Language-focused learning
4 Fluency development

If a course makes use of input that is comprehensible, authentic, and relevant; asks students to produce presentations and participate in discussions about the content; directs attention to vocabulary, structure, and pronunciation features found in the content; and finally, utilizes those components in fluency-building activities, the course will achieve a balance which is effective in teaching speaking and listening skills. The meaning-focused input can be provided by means of readings, audio recordings, videos, etc., as long as it is familiar and interesting to learners, and contains only a small, manageable proportion of language features which are unknown to them (Paramasivam, 2013).

8.6. The safety case

In many circumstances, the curriculum of aviation English courses will want to create the "illusion" that the focus is on acquisition of content rather than on the acquisition of language. Instructors should not shy away from topics that are seemingly complex, as this content is useful, necessary to job performance, and applicable in work settings. By using authentic aviation source materials, technical accuracy can be ensured. Rather than teaching "aviation English," think of it rather as "teaching aviation in English." It is almost guaranteed that most aviation English students are passionate about aviation, intrinsically motivated, and eager to learn more about the subject they are so deeply involved in.

As stated above and in previous chapters, much of the safety content in the aviation world is published in English, but written for expert users of English. This content is therefore inaccessible to much of the NNES aviation world unless an ample amount of supportive scaffolding is provided by curriculum developers and language trainers. In the types of courses described here, content can and should be creatively adapted and utilized as necessary so that learners are receptive to and able to comprehend the information. Adaptations might include slowing down; breaking up, or chunking authentic content; or perhaps rewriting all or portions of the information into a form that is more comprehensible to learners. Other scaffolding techniques include eliciting and utilizing prior knowledge, using content organizers, using transcripts, and simplified paraphrasing (Mathews, 2015).

Awareness and application of safety considerations and practices, obviously, are critical components of any aviation personnel's career competencies. Organizations are known to commit significant resources to improve and manage their safety systems (Mathews, 2014). Therefore, training aviation personnel using safety-relevant content to improve their English language skills has a dual-focus which is attractive to airlines, flight schools, and other aviation training institutions. Within this model, English becomes a tool for learning, acquiring, and sharing new information through safety awareness training which doubles as English language training.

It may not be possible for all safety material or operational training content to be modified for level-appropriate access to learners. However, aviation English can, with language support, essentially provide the mechanism for professionals who need to learn new safety content at all stages of the training. For example, once a curriculum or materials developer has identified content which needs to be a component of operational training, the first step is familiarization with the content. Then, as one example of how to improve the content learning of aviation personnel who use English as a foreign language, the trainer can develop a two- to six-hour (depending on the proficiency of the learners and the complexity of the topic) pre-seminar to familiarize learners with the topic and the relevant vocabulary. Having been "prepped," learners can then complete the training. Access to aviation English instructor support throughout the course will further accelerate their content learning. A debriefing seminar can be offered at the end of the course, where learners might, for example, create a partner presentation on an extension or application of the safety principles or practices and present it to the group. Through this basic sequence, learners acquire the information and use it to practice English skills in both the pre-seminar session and the post-seminar de-briefing with the language trainer. Professionals are expected to learn this information, with or without language support. Organizations that invest in language support resources ensure that their employees are more successful in acquiring and comprehending essential safety information, an increasingly important factor as the complexity of aviation operational systems continues to increase. Enhancing pilots' safety awareness improves their safety culture.

Corpus tools discussed in Chapters 5 and 6 can be of great value to an instructor who intends to utilize elements of authentic safety communication content in the aviation English classroom. Roberts (2018) discusses the use of the corpus tool *WordandPhrase.Info* (Davies, 2018) to identify discipline-specific lexical items from an information-dense text. In aviation English courses, this tool can be utilized by accessing vocabulary classified as "Science/Technology," the discipline that includes aviation/aerospace, which instructors can then use as starting points for the development language of lessons (e.g., pre-reading activities). Instructors who rely solely on intuition to identify technical or operational vocabulary may unintentionally miss important items or include those which are less useful. The identification of frequent and subject-specific vocabulary is beneficial to instructors as they work to build language lessons from authentic content, and the learning of this subject-specific vocabulary is beneficial to language learners as they work to understand content written for expert audiences.

8.7. Putting theory into practice

As noted in previous chapters, especially Chapter 4, the contributions of applied linguistic research to the aviation English classroom are remarkably numerous and useful, particularly with respect to ESP and EOP. In consideration of the complexity of contexts and the specificity of needs, the content and materials utilized in the ESP/EOP classroom will be most suitable if they are tailor-made, so the course developer needs to be capable of developing original material based on both the students´ needs (Bhatia, 2008; Hutchinson & Waters, 1987) and preferences (Kumaravadivelu, 2003). With the objective of providing the aviation English course developer or teacher (often one in the same), especially those new to the field, with strategies and tools to diversify and personalize classes, practical examples are presented here covering applications in various modalities, including classroom-based, online environments (synchronous and asynchronous), or both through blended-learning (ICAO, 2009; Singh, 2003).

8.7.1. Current commercial offerings

To date, there are only a few international publications in the aviation English market. Their importance, however, is unquestionable, and they are a valuable resource for both self-learners and teachers. However, because ESP courses are by nature highly specialized and learner-specific, course design must take into consideration and accommodate a number of important variables, including the background of the learners, training time constraints, the skill of the instructors, and the expectations of the learners (Lin, Wang, & Zhang, 2014). For this reason, the few textbooks which are available may not be sufficient stand-alone sources of materials for courses since on-the-job pilots and ATCs will be expecting more immediately relevant, profession-oriented, tailor-made courses. Their expectations will include, among others, a narrower focus on their specialties, for example, whether they are center, approach, tower, or ground controllers, or even if they are *ab-initio*, general, business, or commercial aviation pilots. It is often necessary to create and utilize in-house, customized teaching materials to fully address the needs of local learners in specific contexts (Paramasivam, 2013). Moder and Halleck (2009) corroborate this view, concluding that professionals do better language work in training activities which are more closely related to their actual work. Unfortunately, however, not all organizations allocate the recommended resources to customized curriculum design and specificity that aviation English courses deserve and, for this reason, these commercially available materials remain widely utilized.

8.7.2. The language needs of on-the-job professionals

- In consideration of the numerous and varied specificities of the job responsibilities of a great variety of professionals in the aviation domain (as mentioned in Chapter 7), it is of utmost importance that any course or materials design begin with a position-specific needs analysis of these professionals while on the job (Hutchinson & Waters, 1987). Such an analysis should be more than simply questioning them about their needs and preferences (Kumaravadivelu, 2003). In the following example, analyses were made *in loco*; that is, data were collected by actually observing professionals (pilots and ATCs) in their work environments (whenever authorized[1]) with special attention to the tasks in which they needed to use the English language. For ease of access, the researchers participated in the daily routines of operations in ATC facilities in Brazil, namely Tower, Approach Control, and Center in different regions. There were also observations of pilots in flight simulators in different training centers in Brazil and in the United States and of "theoretical courses" in airline companies in Brazil. Visits to different airports also helped identify and understand communication needs for ground personnel, particularly during interactions with pilots, and the use of written documents in the flight deck such as operational manuals, logbooks, and checklists. This ethnographic approach identified the kinds of materials used as references and tools, which of these materials are in English, how teamwork occurs in the workplace, and what role hierarchy plays in the workplace culture. This research was done to gain a more comprehensive understanding of the organizational setting and culture as it might influence communication and understanding, and the importance of English in this particular setting. Routine procedures including the recurrent training of personnel were also examined to gain a better understanding of how learning occurs in this context (Henley, 2003). The methodology to gain this comprehensive, data-rich picture included field investigation, shadowing, interviews with supervisors, focal groups and questionnaires with ATCs, course evaluation, and material piloting. The data triangulation disclosed the following realities:

- In the case of ATCs, the three different facilities studied have some commonalities, but also many peculiarities and significant differences concerning language use. Even the dynamics of the job, including the use of body, routines, work design, task components, equipment and other resources, and a number of job elements are very different. Conclusions from this observation and analysis inform the design of pedagogical activities as well as facilitate the selection of appropriate *realia* which most appropriately and accurately relate to the professionals' actual, varied realities when designing courses, lessons, and training activities. While ATCs may share some common aspects in the training protocol, it's apparent from these observations that specific facility/position content is also necessary from one location to another.

- In the case of large countries such as Brazil, the situations/settings, weather conditions, topography, and phraseology vary from region to region and, consequently, location-specific variations are also necessary in linguistic competence training.

- Both ATCs and pilots' training is usually based on simulations of events in order to create scenarios as realistic as possible by confronting them during training with problem-solving situations that help them develop skills essential for high-stakes jobs.

- Pilots reference a large number of English-language manuals in the course of dealing with routine and non-routine situations.

- During stressful urgent or emergent circumstances, pilots and ATCs encounter an increase in both the number of tasks and the speed with which they must be addressed which necessitates concise and unambiguous communications to avoid additional distractions which might escalate the emergency.

8.7.3. Language analysis

English teachers who have not worked in the aviation industry before are likely to be unaware of the characteristics of the language used by the professionals in this domain. To address this need, different projects involving the analysis of primarily aeronautical communications, consistent with the scope of ICAO LPRs (International Civil Aviation Organization (ICAO), 2010) could be designed and utilized. Such a program could be subdivided into sessions utilizing (1) corpora of routine and non-routine situations provided in a particular non-English-speaking country whose ATCs were handling international traffic, or (2) corpora of unusual or urgent/emergent situations handled by international ATCs in different parts of the world. The objective of such investigations would be to identify and describe the language used by pilots and ATCs in both routine and non-routine situations and to analyze and discuss the findings with aviation English students. By examining these corpora in this way, language can be observed empirically in order to build syllabi and materials based on real-life language use for instructional purposes (McEnery & Xiao, 2010; O´Keefe, McCarthy, & Carter, 2007). The corpus of audio files and transcripts may also be reused in the classroom as authentic sources of RTF, as suggested by Paramasivam (2013) and by Prado and Tosqui-Lucks (2017).

The benefits of corpus-based studies and corpus tools are undeniably important in language teaching due to their real-world utility in the selection of language content in both the creation of the syllabus and the design of activities. In this context involving spoken communications, syllabus design needs to include oral language. Friginal (2009) states that "results from the analysis and interpretation of linguistic patterns in a corpus may lead to conclusions about the functional parameters influencing the linguistic choices of speakers" (p. 6). Therefore, the investigation should not only focus on lexico-grammatical aspects, but also on the functions certain patterns of (spoken) expression accomplish in the language domain being studied. By identifying these elements, instructors as well as students can also better describe characteristics resulting from the interaction of the two professionals who are the object of the present research: pilots and ATCs. Prado and Tosqui-Lucks' (2017) analyses of two corpora of ATC communications in routine and non-routine situations and unusual or urgent/emergent situations revealed that:

- Social conventions and perceived hierarchies do play a role in both routine and non-routine communications and are evident in the language used by the professionals, particularly in the use of honorifics such as *sir* and *ma'am*, formulaic expressions and greetings (at times in the local language), as well as politeness markers. This indicates a need to address pragmatic competence.

- Pilots and ATCs do not know all the words in the aeronautical field; however, they use strategies that facilitate their communication in a collaborative way. This means that other than pragmatic competence, strategic competence should be considered.

- Fluency markers or fluency-enhancement strategies help both native and non-native speakers construct and facilitate efficient conversations and are, therefore, necessary components in the sometimes fast-paced problem-solving situations they face together.

- Mitigation of language often occurs, demonstrating a felt need for face work (Goffman, 1967), that is, the necessity of the interaction participants to protect their own image as well as others. This demonstrates the importance of cooperation between and among aviation professionals [see Linde, Goguen, and Devenish (1987) to meet a need for mitigation in the flight deck environment].

- In the transition from radiotelephony to plain English, there are some elements which signal that transition, and allow the professionals to switch the language to a more spontaneous conversation (Garcia, 2016; Lopez, 2013).

- Plain English is usually built through chunks that show politeness or mitigation, or lexico-grammar structures that are routinized in this kind of communication, corroborating the description presented by ICAO Document 9835 (2010) which states that even "plain English [should be] clear, concise and unambiguous" (International Civil Aviation Organization (ICAO), 2010, p. 3.6). To illustrate, the verbs frequently used are either taken from Aeronautical Phraseology (such as *"climb"*, *"descend"*, *"maintain"*) or support verbs (such as *"do"*, *"make"*, *"take"*, *"get"*) in the present continuous, simple past, or present perfect forms; future statements expressing intentions are communicated with both *will* and *going to*; complex structures are not significantly used; subordinate or coordinate sentences are used with fewer conjunctions (*and, but, so, if, because, due to*); questions are commonly embedded after verbs such as say, inform or state (Prado, 2015).

- The type/token ratio revealed a low lexical density, demonstrating a high degree of repetition of patterns and words.

8.8. Real language in language teaching

Based on these discussions above, a curriculum designed for this particular audience should help develop critical thinking, especially with the idea of exploring *native speaker concepts* (Widdowson, 1994); create real-life scenarios and exercises with which students can readily identify; require the use of problem-solving techniques prompting students to focus on the task more than on the linguistic aspects of the scenario; and assist students with constructing language by allowing them to recognize certain linguistic features they are usually unaware of, such as fluency-enhancement strategies, mitigation, etc. (Prado & Tosqui-Lucks, 2017).

Instructional design is informed by theories of language acquisition and language learning pedagogy (Tomlinson, 2009; Tomlinson, 2011). Having adopted the ESP/EOP perspective, which is based upon adult learning in professional environments, pedagogical approaches such as CLT and CBLT are used (see the previous section). Communication utilizing content in this specific domain may commonly require the use of certain supplemental teaching tools, such as aeronautical charts, video simulations, radio communications, and authentic written text sources such as operational manuals, checklists, logbooks, and accident and incident reports as prompts so that the students can produce the target language in relation to meaningful and realistic tasks. The field investigation carried out in the first contact with the professionals identified such *realia* and tasks that could be incorporated into the curriculum.

8.8.1. Blended learning

Time constraints resulting from busy schedules and work shifts are important considerations in the practical design of a model for aviation language training (ICAO, 2009). The alternative to pure classroom-based instruction is blended learning, which is the use of multiple delivery media to accomplish learning outcomes (Singh, 2003). The combination of pedagogical activities such as real-time, face-to-face classes and web-based courses makes it possible to provide alternatives and expand the exposure time to second language input and practice and give students a variety of options to practice skills and increase understanding. The importance of human intervention in the learning process cannot be overlooked or minimized, however; without it language learning may not be as effective. It is the language teacher who will select the content, guide and support the practice exercises, foster or maintain the focus on the target language, assist the students with learning strategies, and promote essential interaction among participants (Soleimani & Esmaili, 2016).

When students are able to explore the online learning environment at their own time and pace, they can reach out to customized asynchronous websites developed by teachers, where they can access authentic content with step-by-step comprehension activities or language awareness exercises. Some teacher-friendly platforms support files such as simulations, 3D animations, game-like activities, and virtual interaction of four types: learner-content, learner-learner, learner-instructor, and learner-interface. Free software such as Moodle[2] or ExeLearning[3] (see Figure 8.1), or even ones available for a fee (many with free trials) such as Blackboard Learning[4] or Adobe eLearning tools,[5] are a few useful examples.

At the same time, some students are more dependent on and successful with an instructor guiding them throughout the activities or, perhaps, even interacting with them as they progress toward the acquisition of communication strategies and skills discussed above. The absence of interaction in asynchronous platforms can be supplemented with synchronous learning environments, either face-to-face in a classroom or online. For the latter, online virtual classroom platforms are recommended when building computer-based activities, because they serve not only as video conference devices but also as tools for note-taking and the manipulation of *realia* (Figure 8.2), consistent with the common visual learning style of pilots and ATCs. Some examples of virtual classrooms are BrainCert,[6] LearnCube,[7] WizIQ[8] for paid versions, or conference tools with add-ons which support interactive boards such as Zoom[9] (free for a limited number of users).

By utilizing such options, students are able to adapt their learning time to their sometimes crowded and irregular schedules, preliminarily explore topics before they are introduced or after as review, proceed through the content at their own pace, or use the website courses as reference whenever needed. More information on blended learning is widely available online, for example, Horn's webpage at: https://www.blendedlearning.org/michael-horn/.

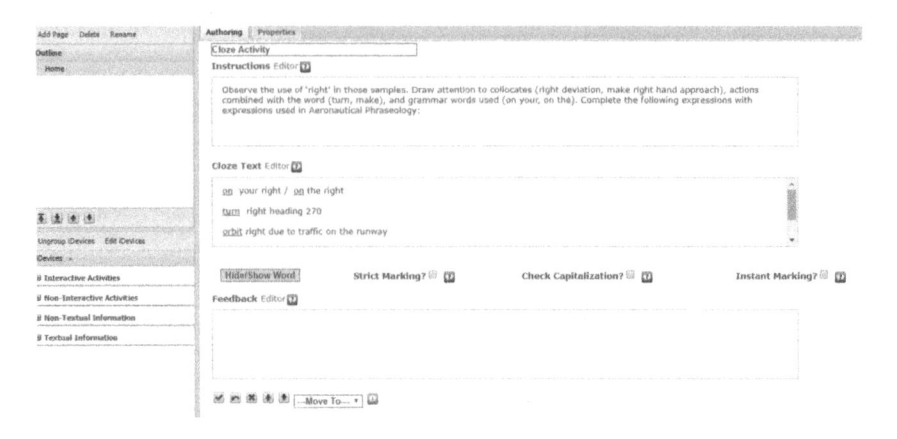

FIGURE 8.1 *Sample cloze test activity under construction.*

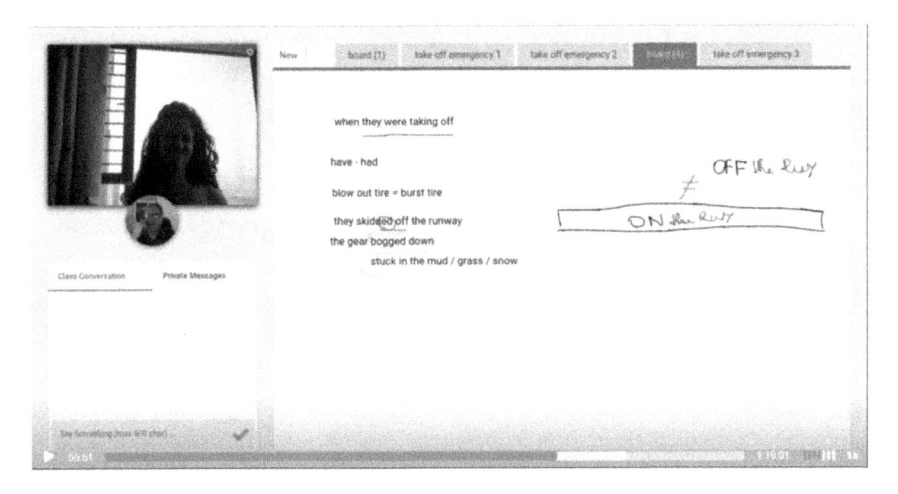

FIGURE 8.2 *An example of a synchronous virtual classroom using an online platform.*

8.8.2. The application of corpora in the classroom

As discussed in Chapters 5 and 6, in addition to benefiting with syllabus design, the use of corpora and corpus tools is also valuable in creating pedagogical activities. Authentic texts are inherently disorganized and, without supplemental facilitation by the instructor in use, may not be user-friendly as a stand-alone resource. [See Carter and McCarthy (2017) and Friginal (2018) for further discussion.] However, research has demonstrated that language learners and professionals find that exposure to authentic audio which presents successful communication between and among aviation professionals is particularly useful as models of language and cultural and situational transactions which they will customarily encounter when doing their jobs (Friginal, 2009). Using the audio files from the corpus allows for this type of exposure and analysis in the classroom, and the language teacher can reference particular aspects of the transactions as they transpire in order to raise and reinforce awareness of the elements of successful communication, including the relevant strategies, protocols, and cultural aspects of those specific communications. In addition to raising awareness of broader features of the discourse, teachers can also focus on particular elements and patterns/lexico-grammar such as chunks which mitigate language or make language more polite, for example, a common characteristic of radio communications. By listening to actual transactions, students will "establish ... the relationship between task performance and linguistic performance" (Friginal, 2009,

p. 291). The teacher may present data such as the most frequent words/chunks in aviation English communications and reinforce the point by referring to examples in different audio files as they actually occur.

8.9. Sample (corpus) lessons

The following exercises are either (1) corpus-driven, that is, they utilize data extracted directly from the corpus, or (2) corpus-informed, that is, they have been prepared in a way that highlights concepts evident from the corpus without using the corpus data directly. These activities are meant to illustrate potential uses of corpus data in aviation English teaching, but may need to be adapted, expanded, or shortened depending on the specific context and needs of learner populations.

8.9.1. Activity 1: Right!

Activity 1: A *right* deviation or the *right* deviation?	
Modality	Group blended learning The first part of the class is online, with exercises guiding the students toward content and language awareness. The discussion will broaden in the second part, when students meet face-to-face with the instructor.
Objectives	Clarifying miscommunication; awareness of strategic competence and fluency-enhancement exponents
Skill	Speaking
Task	Request/approve weather deviations/pre-flight or pre-shift briefings
Aviation Content	Weather deviations, justifying
Language Feature	Ambiguity of language, importance of chunks and collocates
Context	Swinehart (2013) investigated a corpus of cockpit recording transcriptions from the National Transportation Safety Board (NTSB) and concluded that the polysemic word "right" appeared only 18.2 percent of the time with the meaning of direction as prescribed by Aeronautical Phraseology documents. More than 80 percent of its meanings referred to "correct," "exactly," or "just," being used as a discourse marker, a tag question, or a speech act for confirmation. For this reason, it is important to discuss such matters in the aviation English classroom and raise awareness on how to identify potential and/or prevent or resolve ambiguity issues in communication.

DEVELOPMENT OF AVIATION ENGLISH PROGRAMS

Pedagogical Sequence

Part 1: Online, asynchronously before class meeting [Note: Use an appropriate image file to clearly illustrate the situation. Weather maps from various online sources could be selected for this activity].

   a Look at the picture below and think about the following questions:
      1 What would your day be like if this situation happened to you?
      2 What are the biggest problems and challenges a pilot/an ATC can face with this kind of condition?

   b Check the items you believe would be addressed in a briefing (pre-shift or pre-flight) in the case of the scenario above:
      ( ) vertical restrictions
      ( ) deviations
      ( ) holdings
      ( ) alternate airports
      ( ) TCAS maneuvers
      ( ) airport disruptions
      ( ) runway incursion
      ( ) fuel consumption

   c Situations such as the above require much more communication between ATCs and pilots than would be necessary in good weather conditions, particularly when it comes to requests/suggestions concerning deviations. Language research found that the word **"right"** is only used 18 percent of the time in pilot-controller communication to indicate direction, as utilized in Aeronautical Phraseology models.

   d [Note: Provide students with images or extracts of the uses of **right** in aviation texts, for example, in ICAO *Document 9432: Manual of Radiotelephony*]. Observe the use of **"right"** in those samples. Draw attention to collocates (e.g., *right deviation, make right hand approach*), actions combined with the word (*turn, make*), and structure words used (*on your, on the*). Complete the following expressions with expressions as they would be communicated in Aeronautical Phraseology:

<u>on</u> your right / <u>on</u> the right
<u>turn</u> right heading 270
<u>orbit</u> right due to traffic on the runway
<u>take</u> first right
request <u>right turn</u> when airborne

e Read the following communications:
- *Control / Airplane247 //*
- *Airplane247 / go ahead / sir //*
- *Airplane247 / crossed POSITION1 on 215 / FL390 / estimating POSITION2 12:23 / POSITION3 is next / squawk 6671 //*
- *Confirm / are you making deviation to the right of of of the route? / confirm? //*
- *Say again? / I'm sorry / Airplane247? //*
- *Airplane247 / confirm / are you making deviation to the right? / confirm //*
- *Ah / negative / no / we / we can remain on course // ah / do you need us to change the altitude to correct our flight? //*
- *Roger / Airplane247 //*

In the communication transactions above, consider and respond to the following questions:

1 Why did the ATC ask about the deviation? What do you think happened?

2 Why do you think the pilots didn't understand the question?

3 There was some hesitation in the pilot's utterance (*ah / negative / no / we / we can remain on course*). When do hesitations like these occur? Why?

4 How did the pilot try to make sense of the apparent confusion?

f Observe the items below, which were all taken from actual communications, and identify:
- An expression that means "it's correct"
- A way of requesting deviation
- A way of approving deviation
- Two occurrences of request for immediate action
- A structure word that usually combines with "the right"

   1 *Seven, confirm, are you making deviation to the right? Confirm.*
   2 *Start to make hold right now*
   3 *Start hold right now/AIRPLANE/Confirm/turn right/confirm?*
   4 *We cannot turn to the right//*
   5 *… that you'll not be cleared to descend right now/you have to wait …*

6 ... the present position to the right/confirm?///AIRPLANE/To the left//

7 You are under/radar vectors right now/maintain the heading two seven

8 AIRPLANE/cleared right deviation after deviation fly direct POSITION//

9 AIRPLANE/requesting right deviation due to weather/heading 080//

10 AIRPLANE/that´s right and report passing position XXXX please//

11 leave hold now/turn right heading POSITION//

12 make hold over POSITION with right turns

Part 2: In class asynchronously, face-to-face, in groups or one-on-one

  a As you've seen in the online lesson, the word **"right"** is considered ambiguous.

    a Which expressions do you think could cause misunderstandings and why?

    b How would you fix them?

    c Can you think of other words that have caused confusions?

    d What words can cause confusions in radio communications?

  b [Note: Preferably in groups] Refer to the satellite image again and think of five possible sentences/utterances you can use to request deviations/to approve deviations using phraseology. Try to consider words that might cause confusion (from Exercise A).

  c Present and discuss your ideas with the group/the teacher.

  d Listen to a group of students discussing the questions above. Did you reach the same conclusions?

8.9.2. Activity 2: **Wrong!**

Activity 2: Wrong on the Spot	
Modality	One-on-one (face-to-face or online synchronous)
Objectives	Placing aviation English as a shared language among speakers of a professional community (and improving the self-image of the international pilot)
Skill	Speaking/Listening

(Continued)

	Activity 2: Wrong on the Spot
Task	Briefing/discussion on hotspots/analysis of a communication
Aviation Content	Runway incursion/airport layouts/ATC-pilot communication
Language Feature	Transition from Aeronautical Phraseology to plain language
Context	An airline pilot, foreign in that territory, made a mistake easily corrected in the pilot-controller interaction in the main airport in Miami, USA. Their communication spread in social media, alleging poor English. The intention is to spark a discussion on whether or not there was in fact communication failure.

Pedagogical Sequence

The following sequence is written as directed to the teacher in an asynchronous class. This activity could be done either in an online virtual classroom or a physical in-person classroom. While interaction in groups is encouraged, this lesson can also be delivered one-on-one with a single student.

a Present the Miami International Airport's chart[10] (ICAO code KMIA) (see Figure 8.3). Ask students to conduct a briefing on the airport layout, highlighting the four hotspots on the chart. (A hotspot is defined as "a location on an aerodrome movement area with a history or potential risk of collision or runway incursion, and where heightened attention by pilots/drivers is necessary" (ICAO, 2007b)). Discuss the reasons why those areas are considered hotspots.

b Present the communication audio, preferably from KMIA, and distribute the transcription below. Ask students to use the chart to locate the aircraft on the map, based on the communication audio. Refer to the communication, drawing attention to the readback of the pilot, which demonstrates precise understanding of the clearance.

Hotspot Confusion

ACFT 8043 heavy / Miami Tower / runway two seven / line up and wait / caution wake turbulence departure boeing seven four seven //
 Can you confirm / ACFT 8043? //
 For you ACFT 8043 heavy / Miami Tower / runway two seven / line up and wait //
 Two seven / line up and wait / ACFT 8043 //

DEVELOPMENT OF AVIATION ENGLISH PROGRAMS 241

ACFT 8043 heavy / wind three six zero at one four / gust two zero / fly heading two seven zero / runway two seven cleared for take-off //
Fly heading runway heading / ACFT 8043 //
ACFT 8043 / sir you´re <break> you´re joining the wrong runway / uh it was runway two seven / cancel take-off clearance / hold your position //
Sorry sir / my mistake / I made a mistake / sorry / sorry sorry //
ACFT 8043 heavy / join runway three zero / backtaxi down runway three zero / turn right on tango //
Okay / join uh three zero / join tango to holding point two seven //
ACFT 8043 heavy / continue down runway three zero / turn right on tango and join taxiway Quebec //
Okay / join tango / and I take Quebec / ACFT 8043 //
ACFT 8043 heavy / caution wake turbulence runway two seven / line up and wait //
Okay / runway two seven / caution wake turbulence / line up and wait / ACFT 8043 //
ACFT 8043 heavy / fly heading two seven zero / runway two seven / cleared for take-off //
Fly heading one <break> two seven zero / runway two seven / cleared for take-off / ACFT 8043 //

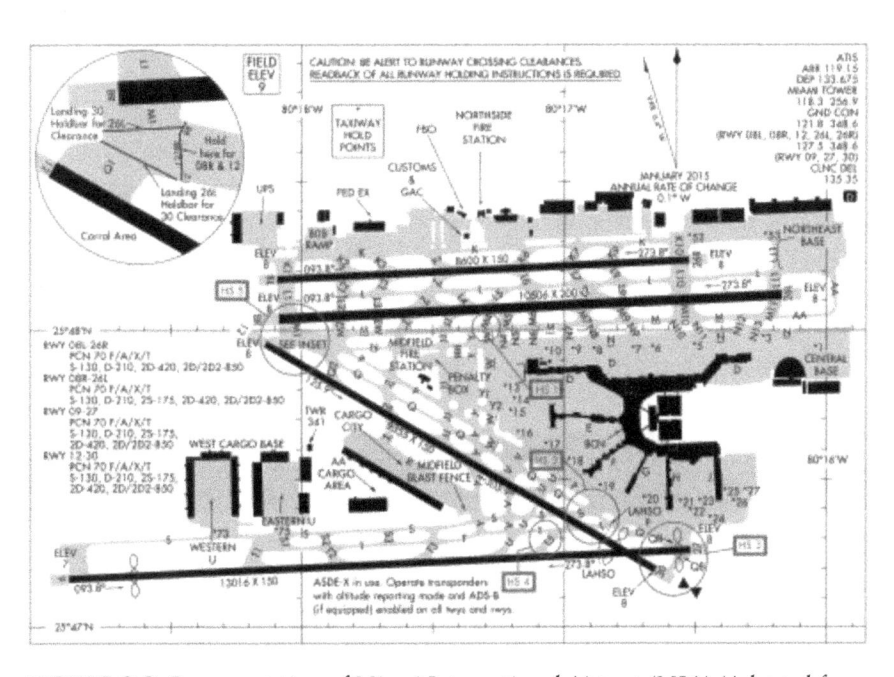

FIGURE 8.3 *Representation of Miami International Airport (MIA) (Adapted for this book from SkyVector.com).*

c Present the following questions to students and have them discuss in groups:
   1 What kind of violation happened in this event?
   2 Was there any communication issue? Did the pilot or the ATC demonstrate any communication difficulty, especially when trying to comprehend each other? Why (not)?
   3 Does the tone of the Tower controller's voice show irritability or intention to punish the pilot? Why (not)?
   4 What was the pilot's reaction? What did he say?
   5 Was there cooperation between the two professionals? What was the outcome?

d Raise the following questions to students and have them discuss in groups:
   1 What kind of violation happened in this event?
   2 Was there any negative communication issue? Did the pilot or the ATC demonstrate any communication difficulty, especially when trying to comprehend each other? Why (not)?
   3 Does the tone of the Tower controller's voice show irritability or intention to punish the pilot? Why (not)?
   4 What was the pilot´s reaction? What did he say?
   5 Was there cooperation between the two professionals? What was the outcome?

e Next, show the concordance table below (Table 8.1) which contains information from the corpus to illustrate the use of the word "sir" when signaling the transition from phraseology to plain English. Refer to the audio file or transcript again and ask the students to compare the communication to the corpus data. Question the reason for the use of the word "sir" both, allowing students to brainstorm, report, and confirm the instances in which these chunks might be employed.

Table 8.1 Concordance lines for *sir* from a corpus of aviation interactions

In RTF communications, *sir* is one of the top 10 words. Some of the most common expressions are …		
Yes, sir	Thank you, sir	Sir, uh

N	Concordance		
1	ir//Alright/we'll wait/	sir	//That was for somebody el
2	two three four?//Go ahead	sir	//We'd like to divert uh b
3	inue climb two seven zero//	sir	uh/cleared to climb level
4	ndred/I understand?//Yes	sir	//Cactus seven thirty-nine
5	er now/thanks for the help	sir	//Delta ten sixty-three we

In RTF communications, *sir* is one of the top 10 words. Some of the most common expressions are ...		
6 ency due to a bird strike//	**sir**	/Delta ten sixty-three?//
7 cleared to land//Thanks/	**sir**	//Nippon Cargo one ninety-
8 ghty-five/Say again//Yes	**sir**	/this is United eight eigh
9 uple of fatalities//Okay/	**sir**	/I'll let them know/San
10 again//Yeah three sixty/	**sir**	/just check we are on nine
11 ham//Just for our planning	**sir**	how long is six/two-four go
12 ing to seven thousand feet/	**sir**	//we have flown by a ballo
13 alloon now on the right/uh	**sir**	there's plenty of balloons
14 Europa zero five seven//uh	**sir**	we are now turning uh right
15 lpha that's not your traffic	**sir**	you were just supposed to f

f After the discussion of the communication, ask students to think about their own experiences. Have they ever gone through any similar event? Could such an event have happened to them? What actions can be taken to avoid such problems? If they happen anyway, what would some proper actions/responses be?

g Next, raise questions and potential issues related to the identity of the NNES coping with aeronautical communications such as these. Facilitate a discussion concerning the fact that this incident was spread over social media as an example of poor proficiency in English, even though it didn't happen as a result of language problems. Ask how students feel when flying in international airspaces or when dealing with international traffic when the other participant is a native speaker of English ... and the reasons why they feel as they do in such situations.

8.10. Concluding remarks

Aviation English can be regarded as being under the umbrella of ESP; however, the aviation English teacher needs to be equipped with a broad understanding of the aviation industry and its operations, which requires more in-depth industry knowledge and focused attention on specific job contexts and knowledge of the responsibilities and functions of each job classification. That means the aviation English teacher usually works only within this domain, as achieving even the minimum knowledge and expertise essential to do the job adequately can take longer than to acquire the corresponding abilities necessary to teach general English. In fact, ICAO considers up to three years of experience to be necessary to reach this level of qualifications (ICAO, 2009).

The work of developing aviation English curriculum and materials requires a great deal of attention to detail and awareness of operational considerations. The lack of context-specific commercially available materials means that aviation English classrooms are often replete with activities and materials designed in-house to fit the student population's profession, operational and language skill level, and availability. The tools presented in this chapter and the accompanying discussions are an attempt to support and facilitate the work of the aviation English teacher in the classroom. The theoretical framework, rationale, and research supporting the suggested pedagogical approaches with the various occupational populations within the aviation industry have been briefly presented together with the suggestions and recommendations themselves.

The final section of this chapter presented two sample lessons illustrative of the kinds of curricular approaches that can be utilized in training in the aviation industry, one face-to-face or online synchronous class, initially designed for pilots and the second one focusing especially on ATCs in a blended or hybrid format, that is, partially delivered asynchronously through web-based tools and partially delivered face-to-face/online with web conference tools. Both lessons are well-suited for groups of professionals, especially since these two positions are typically interacting routinely in the normal work situation. The focus in this section is on-the-job professionals whereas, in contrast, the following chapter will focus on and provide similar practical suggestions specific to *ab-initio* pilots. The primary intent here—as well as throughout this book—is not to mandate "recipes" or comprehensive, prescriptive descriptions or rules of what to do in the aviation English classroom and how to do it, but rather to present guidance, examples, information, and the rationale and research foundations, when available, upon which these resources and recommendations are based. A critical component of these lessons is that they should be tailor-made to fit the context at hand, and an important role of the aviation language instructor is to assess that context and develop plans and approaches to meet the needs accordingly.

Chapter Co-Authors

Malila Prado has been teaching aviation English to pilots and to language trainers in Brazil since 2008. As such, she has designed in-house aviation English material to flight dispatchers, flight attendants, and pilots. She has been involved with research of aeronautical communications in the fields of corpus linguistics, language teaching, and pragmatics. She holds an MA degree in English and is currently a PhD student at the University of Sao Paulo, Brazil. She is a member of two aviation English research groups: ICAEA-RG

(International Civil Aviation English Association Research Group) and GEIA (Aviation English Study Group—Brazil).

Patricia Tosqui-Lucks holds a PhD in Linguistics, focusing on teaching English for Specific Purposes. With over twenty years of experience teaching English in various contexts, she has been the supervisor of the Aviation English training program for air traffic controllers at Brazil's Airspace Control Institute since 2009. She is also the leader of GEIA, an Aviation English research group in Brazil, with published articles and books in the field. Currently, she's conducting post-doctoral research at University of São Paulo, in Brazil, about applying corpus linguistics to Aviation English course design and material development. She's a member of ICAEA-RG.

9

Ab-Initio Aviation English

9.1. Introduction

This chapter outlines three aviation English courses, all ESP-based, which were created and implemented at Embry-Riddle Aeronautical University. All designed for the same overarching purpose, these courses aim to prepare *ab-initio* aviation students for their future flight training. However, all three courses are intended for slightly different contexts and, accordingly, they vary in their balance of content use and language focus. A testament to the value of tailor-made, context-driven aviation English curriculum development, these courses represent a shift from the "one size fits all" characteristics of commercial aviation English teaching materials. As one might expect, given the great variety of aviation English contexts outlined in Chapter 7, the few commercially available textbooks and programs simply cannot be expected to meet the needs of all fields within the aviation industry which employ NNES personnel, or to provide appropriate training at all points in a trainee's language and content development during their career.

As the need to train more NNES pilots grows exponentially each year, so does the demand for programs, which enable flight training to progress in a manner which is cost-efficient, timely, and above all, ensures safety. Furthermore, there is the need for novice air traffic controllers, maintenance technicians, etc., to be taught foundational English language skills to enable them to participate successfully in their EMI training. While focusing primarily on *ab-initio* pilot training, these courses which utilize foundational topics in aviation to teach language skills necessary for training environments can also prove useful for those pursuing other careers as well. The majority of this discussion, however, will focus on the utilization of these materials to prepare individuals for flight training.

Despite ICAO LPRs not referencing *ab-initio* pilot training, the FAA policy, for example, for Aviation English Language Standards (AELS) requires cadets to be at ICAO Operational Level 4 to receive the "English Proficient" endorsement on FAA licenses. In addition to helping individuals be better prepared for flight training, these *ab-initio* aviation English courses may also help prevent pilots lacking requisite proficiency from receiving an English Proficient endorsement due to evaluation by a non-language specialist, and even more importantly, these tools may help make flight training more safe and efficient.

All ESP courses, including aviation English courses, should be designed with a predominant learner-centered approach, providing students with engaging learning experiences that enable them not only to learn content, but also to learn skills to use the language associated with the content in different contexts (Aiguo, 2008). It is important to note that in the flight training context, once students finish their aviation English course, they will begin what is termed "ground school," which is, typically, a learning environment characterized primarily by lectures imparting great amounts of information and terminology augmented with PowerPoint presentations, similar to courses typically found in universities. To receive and understand the information, students need strong listening comprehension skills and the ability to formulate questions, participate in discussions, and read text saturated with subject-specific vocabulary. The operational side of training introduces new language skills challenges, such as interacting with a flight instructor and communicating on the radio. Considering the time and financial investment in providing flight training, and, most importantly, the possibility of a safety incident related to poor English (Emery, 2015), it is of vital importance that students enter flight training with adequate language proficiency required.

The three courses described in this chapter were developed in an effort to mitigate if not eliminate the consequences of inadequate language proficiency upon the effective outcomes of individuals' flight training. Briefly, the three ESP courses in this chapter can be described as (1) Aviation Topics, a theme-based CBI course in an EAP setting; (2) English for Visual Flight Rules, a tandem-taught CLIL course focused on phraseology; and (3) English for Flight Training, an intensive course to prepare NNES for EMI flight training in the United States.

9.2. Aviation Topics course

Intensive English Programs (IEPs) in the United States share a common mission of preparing international students for successful university study by way of developing their English language proficiency, cultural competence, and academic study skills. Students typically take courses including writing, reading, listening, speaking, and grammar, utilizing lots of different reading and lecture materials on a variety of topics, such as the environment, technology, American culture, history, music, art, or psychology. Often, these materials have been written or recorded specifically for use in an ESL classroom. The pace may be slower and the material may contain simplified syntax, lexis, and patterns of organization to facilitate learner comprehension. At lower skill levels, such simplification is necessary. However, as learners progress through an IEP, materials should, ideally, increasingly approximate the types of discourse and levels of difficulty which they will encounter in university study.

Once they have completed an IEP program, students often venture forth across the United States and the world to pursue various degrees ranging from literature to business to engineering. The IEP at Embry-Riddle is somewhat unique in that the majority of its students share a common goal of pursuing careers in the aviation industry, and it is quite common for them to matriculate directly into Embry-Riddle for degrees such as Aeronautical Science, Aerospace Engineering, Aviation Maintenance Science, or Air Traffic Management. With such a large population of students interested in the same industry, the IEP at Embry-Riddle decided to create a new course: Aviation Topics. The development and implementation of this course was accomplished through careful and systematic design, and was funded through a Scholarship of Teaching and Learning (SoTL) internal grant. Prior to implementation, the potential effectiveness of such a course was considered and researched to answer such questions as:

1 Can students achieve the same academic outcomes as other IEP classes (e.g., skills for paraphrasing, taking notes from a lecture, giving a presentation, etc.) using a common content focus and context (aviation) anchored throughout an entire course?

2 What are the effects of preparing students in an IEP course for study at a specific university, that is, Embry-Riddle, through focused instruction with a conscious intention to bring Embry-Riddle into the classroom and, subsequently, the IEP students into Embry-Riddle?

Answers to the first question were determined quantitatively by the scores of a subject-specific diagnostic and final exam. Answers to the second question were determined by a pre- and post-survey administered to the students. The results and analysis are discussed below. Note that, although this course exists currently within an IEP at a university, much of the content and many of the activities are relevant to flight training settings outside of universities, used as-is or with modification.

9.2.1. Goals and objectives

While there is certainly a place for courses dedicated to each individual context for novice aviation professionals, it can be argued that a curriculum can productively cross over and serve a variety of students by using foundational aviation topics as the content. Even outside the university, the foundation of pilot and ATC training (i.e., ground school or basic training) will include a wide range of similar topics due to the close interrelationship within the aviation environment (in Emery, Entry Point North, 2015). Because the course Aviation Topics fits into the wider IEP curriculum, its outcomes necessarily cover not only listening and speaking skills, as an *ab-initio* training course which is part of a flight training institution would be expected to do, but also reading and writing. Many students are preparing for not only flight training, but also academic study. Interestingly, a substantial number of students in Embry-Riddle's IEP are not intending to pursue a course of general academic study, but, rather, intend to enter a flight training program at one of the many schools in the surrounding area. Regardless of subsequent interests and pursuits, aviation is an immediate common interest, fueled for many by a lifelong passion for flight and all things airplane.

The goals for the Aviation Topics course include the following:

- Students completing Aviation Topics should communicate in academic and professional contexts about aviation-related topics.
- Students completing Aviation Topics should take complete notes from authentic sources about aviation that contain all the important/relevant info from the lecture/material.
- Students completing Aviation Topics should be able to understand and utilize authentic academic-level reading materials about aviation-related topics.
- Students completing Aviation Topics should be able to produce written texts in a variety of formats, including essays, written responses, and reports, utilizing authentic academic-level texts in aviation.

In addition to the academic, content-based goals of the class, another goal is to increase students' own positive perception of their preparedness for entrance into Embry-Riddle or other institutions and nurture their excitement to matriculate into the university. Students in IEP programs may feel disconnected from the larger university. In a situation such as Embry-Riddle, where so many students would prefer to be sitting in a flight deck instead of a classroom, the design of this course intends to facilitate opportunities for students to begin experiencing the aviation world more directly and more quickly.

Conceptual underpinnings. As this course exists within a larger IEP program, its broadest goal is the acquisition of language rather than content which is true in other courses in the program as well. However, in the acquisition of language skills, only authentic aviation content is used in the course, without adaptation. A theme-based approach to CBI is used, with the topics ranging from aircraft systems to aerodynamics to human factors to current events in aviation.

9.2.2. Syllabus design

Choosing the content. As iterated previously, the content used is most importantly relevant, interesting, and chosen purposefully to maintain interest and excitement while also immersing students in the world of aviation and, of course, accomplishing essential English language proficiency and competence. Since the students of Aviation Topics will enter either into the university or into a flight training program following the course, the content chosen should also provide a foundation for their future studies. It is almost as though the content from Aviation Topics serves as a placeholder for typical "pre-reading" or "pre-listening" activities; this time, however, the actual reading and listening will take place outside the IEP. In order to prepare for this course, the course developer took an Embry-Riddle course, Principles of Aeronautical Science, whose course description is as follows:

> An introductory course in Aeronautical Science designed to provide the student with a broad-based aviation orientation in flight-related areas ... [including] historical developments in aviation and the airline industry, theory of flight, airport operations, aircraft systems and performance, elements of air navigation, basic meteorology theory, air traffic principles, flight physiology, and aviation regulations and safety.

This course is likely one of the first courses that Embry-Riddle students will take once matriculated and covers a range of foundational topics for anyone entering the aviation world. Through participating in Principles of Aeronautical Science, the course designer acquired core aviation knowledge, which subsequently formed the basis of the content for Aviation Topics. The consistent focus on only aviation topics throughout the course allows students to continuously build on their subject knowledge of aviation, eventually transitioning further into increasingly complex aviation topics requiring more critical thinking skills, such as the looming aviation personnel shortage, crew resource management, and even environmental concerns in aerospace engineering. No specially contrived readings or lectures are used in this course; with an appropriate amount of scaffolding, students can comprehend, analyze, and interact with authentic content from real-world sources, such as those from Principles of Aeronautical Science. In fact, the use of authentic content intrinsically motivates students who are passionate about the subject matter, thereby enhancing the student learning experience by enabling students to see that they are ready for this level of content.

Additionally, students are introduced to a great deal of general university-specific knowledge as they progress in the course, gaining familiarity with the physical campus through class visits and tours, communicating directly with professors, and utilizing materials from other university courses in the IEP classroom.

Designing the tasks. The pedagogic tasks in Aviation Topics correspond to those generally employed in the classroom environment, for example, listening to lectures or videos and demonstrating comprehension, participating in discussions, negotiating and collaborating to produce group projects, and researching deeper into subjects for oral presentations. For example, at times university professors are invited into the classroom to give full-length lectures. Furthermore, the course developer compiled information from authentic sources and created lectures which are up to one hour in length and delivered by the instructor. Through these experiences and learning tasks, students emerge from the course having worked on the same skills as their other IEP courses, but having acquired these skills within the context of the specific content area they will likely encounter in the university subsequently. The course designer verified this "realism" of the target tasks utilized in the course by means of participation in Embry-Riddle courses as described above.

Instructional materials. All materials in Aviation Topics were tailor-made by the course designer to ensure authenticity of content and representation of the target tasks in aviation classrooms. Examples of the materials include graphic organizers and listening guides to accompany readings and lectures, PowerPoint presentations utilized in lectures given by the instructor, worksheets to reinforce instructional content and facilitate discussions, as well as rubrics for assessment. An additional document containing detailed day-by-day lesson plans was created by the designer to guide future instructors.

9.2.3. Activity types

Goal one: Language acquisition. Activities which focus on the development of language skills utilizing aviation content employ an integrated skills approach to cover as much as possible reading, writing, listening, and speaking in a single thematic unit. While the activities in Aviation Topics did lean more toward a listening and speaking focus, reading and writing skills were included not only because of the course's position in the IEP program, but also because aviation personnel do have a need for these skills in addition to oral communication. Ways in which aviation personnel use reading and writing skills, in addition to listening and speaking, are covered in previous chapters of this book. Examples of specific activities are described below.

Solar Impulse project. In 2016, Solar Impulse became the world's first purely solar-powered aircraft to fly 40,000 km around the world in an effort to promote clean technologies (Solar Impulse Foundation, 2018). Two pilots flew without any fuel or polluting emissions, powered only by the sun, for five days and nights, a remarkable story with a breadth of interesting and relevant content to be explored by aviation students.

Lesson particulars. The website detailing the Solar Impulse aircraft (https://solarimpulse.com/) is full of information on the vision, the challenges, and the team of pilots. After watching an introductory YouTube video on the aircraft, students engage in a facilitated discussion on how such a feat might be accomplished and the immediate challenges they see in completing it. Students are assigned an overview article to read and summarize for homework from Newsela (2016). [Note: Newsela publishes adapted versions of current news articles for English language learners.] The following day, each student is given a challenge, either human or technical, listed on the Solar Impulse website. Each handout contains several paragraphs on the challenge as well as a link to a YouTube video detailing the challenge. Fifteen minutes are provided to learn about the challenge and write a summary. Students then participate in a gap-fill activity in which they rotate around the classroom, speaking for five minutes to a partner to both deliver a summary of their particular challenge and record information from the partner on theirs. Finally, students are mixed into groups of three to create a final oral presentation on the challenges of Solar Impulse. In this lesson, students then read (the Newsela article and the paragraphs from the website), write (a short summary of their paragraphs and YouTube video), listen (to YouTube videos and their classmates reporting of information), and speak (in class discussions, group work, and final oral presentations).

Crew Resource Management (CRM) project. In addition to communication with air traffic controllers, pilots also need skills for cross-cultural communication on the flight deck. Crew Resource Management (CRM), defined as "the effective utilization and management of all resources—information, equipment, and people—to achieve safe and efficient flight operations" (Pettitt, 1995), was developed primarily in response to accidents attributed to human error and has since become a standard in aviation training programs. It is important that CRM training adapts to the culture in which it is being taught. Hofstede's (1991) seminal work identified four dimensions of classification for culture, and subsequent scholars have researched the application of these aspects to flight crew behavior. The three main aspects of culture relevant to flight crew behavior (also referenced in earlier chapters of this book) are the ways different cultures deal with uncertainty (uncertainty avoidance), the extent to which individuals' behaviors are influenced and defined by others (individualism-collectivism), and the relationship between subordinates and superiors (power distance). The ideal flight crew has strong team orientation (collectivism), easy communication from subordinates to superiors (low power distance), and a regard for rules and procedures (high uncertainty avoidance). Because different cultures will fall in different places along the continuum with respect to each of these dimensions, CRM training should be tailored to fit particular cultures (Engle, 2000; Helmreich & Merritt, 1996).

Lesson particulars. Students are introduced to the topic of Human Factors through a lecture which introduces key information related to broad concepts and then a narrower focus on language and culture in Human Factors research. This discussion leads to an overview of CRM and the likelihood of multicultural flight decks in an increasingly globalized aviation world. Then, ideally, students are partnered with someone from a culture different than their own. The assignment is as follows:

1 Working with your partner, research the cultural dimensions (UA, PD, IND) of your native countries using the "Compare Countries" component of the Hofstede Insights website (https://www.hofstede-insights.com/).
2 Determine what would need to be focused on in CRM training for each country to reach the "ideal cultural profile" of a flight deck.
3 Consider a flight crew made up of individuals from your two countries and Americans. What would work well? What would be specific challenges?
4 Write a report which answers each of the questions above, and be prepared to orally present the information to the class.

In this assignment, a variety of skills are practiced due to the collaborative nature of the assignment, as well as the challenge of researching information and thinking critically to create a report covering the findings for presentation. Again, there are aspects of reading, writing, listening, and speaking present in this lesson.

Goal two: Perception of preparedness. In addition to activities which support the development of language and content knowledge, certain activities aim to target the necessary skills for building students' confidence and positive perception of their readiness for study in the university. These activities, therefore, aim to take students outside of the IEP bubble and into the university setting and allow them to experience moving confidently and comfortably in the wider academic community.

Observing an Embry-Riddle class. Students are required to observe one Embry-Riddle class, take notes, and summarize their experience. On their own, they research information about the class, look up details about the professor, and email to ask for permission to attend and observe the class. Three separate grades are given for this project: researching the class/professor, emailing the professor, and summarizing the experience.

Lesson particulars. Prior to the beginning of the semester, the instructor is responsible for identifying potential classes for students to observe and obtaining professors' preliminary consent and support. Qualifying courses need to be 100 or 200 level and held during the same time slot as Aviation Topics, so as not to interfere with other IEP courses. Students are then provided with a list of prospective courses to observe. They are given the names of the professor, the course number, and the course title. It is their job to research information about the professor, such as their college, department, and official title, and the time and location of the course. This research builds student familiarity with the university's course management system as well as its internal websites detailing college information. Students turn in this information, along with an explanation of why they want to observe their top three choices and, on a first-come-first-serve basis, students are assigned classes according to their preferences.

Once students are assigned a course, they must then email the professor. Students are provided guidelines on how to properly compose an email to a professor and are required to ask for permission to observe through this email. Finally, students observe the course, take notes, and summarize their experience. Representative student feedback is as follows:

> *Last Wednesday, I observed one class at Embry Riddle Aeronautical University called Macroeconomics in order to improve my listening skill and knowledge. This course was explained by the professor who is an excellent teacher. Firstly, I had some difficulties to understand and to listen because the professor was using a lot of academic words and it was my first time to listen to him. But, after twenty minutes, I began to listen and understand what he was teaching.*
>
> *I wasn't expecting that I will like this experiment ... I learned many things in this lecture, and I wish if I can observe this class again. I used to watch in YouTube education movies or some episodes about Robots and unmanned aircrafts, but it was first time to observe lecture about this subject. The professor was well prepared and he has a good back ground about this kind of operation. Moreover, I had chit-chat with him after the class, and I found*

out he is a pilot and he still flies aircraft. I really like the idea of observing classes, and I will start to request to observe more classes. Thank you.

In addition to observing within classrooms, students are invited to observe a class in the flight deck for extra credit. After taking an observation flight with a current student pilot and a Certified Flight Instructor (CFI), students answer questions regarding the maneuvers practiced and the communication heard during the flight. This observation flight takes place after an explanation of ICAO LPRs, and students use this opportunity to observe the impact of pronunciation, structure, fluency, etc., on a "real-world" flight situation.

Guest lectures. A common, important goal of IEP listening courses is to strengthen students' ability to listen, comprehend, and take notes in academic lectures. To this end, hundreds of textbooks exist with accompanying audio or audiovisual lectures which are recorded for the purpose of aiding ESL students in developing listening skills. The problem with using these available lectures in a class such as Aviation Topics is (a) there are fewer lectures with appropriate and accurate aviation content, and (b) the lectures are not authentic materials, which are inconsistent with the stated goal of the course to use only authentic aviation content.

Lesson particulars. To solve both the content and the authenticity issues of most ESL textbook lectures, professors from the university were invited into the classroom to give lectures on the same material and in the same manner that they deliver the content in their classrooms in the university. Aviation Topics students heard one-hour lectures on topics such as the history of aviation, branches of aerospace engineering, accident investigation, and language as a human factor. During the lectures, students took notes and later completed summarizing, presentation, or further research activities based on the information from the lectures. In this way, students reported feeling more connected to the university and having a better idea of what to expect once matriculated.

Facility tours. The location of the university made it possible for the course designer to organize a trip to the local airfield following lessons on airport signage, lighting, and markings. In addition, guided tours of the Applied Aviation Sciences labs were arranged, including air traffic control and meteorology training rooms.

9.2.4. Results and observations

One diagnostic and one final examination were given to students to assess their language skills in an aviation context, requiring skills in both English proficiency and aviation content knowledge. In addition, students were questioned to determine their current perception of aviation knowledge and their level of preparedness for matriculation into Embry-Riddle. The questions are as follows:

1 I have knowledge about aviation.

Strongly Disagree		Somewhat Disagree		Neutral		Somewhat Agree		Strongly Agree
1	2	3	4	5	6	7	8	9

2 What type of experience do you have learning about aviation?
3 I am prepared to enter Embry-Riddle Aeronautical University.

Strongly Disagree		Somewhat Disagree		Neutral		Somewhat Agree		Strongly Agree
1	2	3	4	5	6	7	8	9

Aviation knowledge. In the initial diagnostic exam, the average score across two semesters for question one, aviation knowledge, was 3.26. In the end-of-semesters final exams, average score across two semesters for the same question was 7.70. The average score on the diagnostic exam was 13.5 percent; the average score on the final exam was 85.3 percent.

Embry-Riddle preparedness. In the initial diagnostic exam, the average score across two semesters for Question 3, Embry-Riddle preparedness, was 3.30. For the end-of-semesters final exams, average score across two semesters for the same question was 8.05. Aviation Topics continues to be a regular course offering in the Embry-Riddle Language Institute, providing future pilots, air traffic controllers, maintenance technicians, aerospace engineers, and even business professionals in aviation-related positions with specific preparation for their careers, and giving them greater familiarity with the aviation content that is part of the context within which they will work and learn.

9.3. English for Flight Training course

English for Flight Training is a course for flight training students who have not yet begun their actual flight training. This is a four-week course, meeting Monday through Friday for six hours each day. Flight school programs, as well as their students, have benefitted from this course by offering it the month before a new cohort of *ab-initio* students begins ground school. In the same way that university IEPs prepare students to enter academic study, the English for Flight Training program prepares students for their entrance into a flight training program. With a total of 120 contact hours within one month, the course is certainly considered as intensive as a university IEP which must offer a minimum of eighteen contact hours of English language instruction per week to be considered intensive (EnglishUSA, 2018). ICAO (2009) recommends this intensive approach as it offers "greater scope for continuity, recycling activities for consolidation, and enhancing the quality of tutoring and of the student-trainer relationship" (2.3.2, p. 17).

Although the course of study was designed and intended for a comparatively specific profile of learners, the profiles of students actually enrolling and demonstrating measurable improvement have, thus far, been much larger and more diverse than was originally expected. The following is a description of the most common types of students. The first one is the intended population targeted by the course design:

1. A student who has already enrolled in flight school and will begin in that program immediately upon completion of this course. This student may have some aviation content knowledge, but likely in the first language only. Because this student is typically young and coming straight from high school to the EMI flight school, there may be aviation content knowledge from a range of training and other experience in the home country, including time using a simulator, often in live online role-plays, time learning about aviation through media, and even time in classes to prepare for an aviation career. This student may have high English language ability.

2. A NNES pilot who has never trained in an EMI environment and does not fly international routes. This student likely has low ICAO language proficiency, although this may not always be correctly represented by license endorsement or evidenced by previous ICAO language proficiency exams. This student will likely have strong content knowledge in aviation, but with ability only in the first language. Students in this category have been observed to perform the most strongly in radiotelephony activities.

3. A student who begins flight training, but becomes "grounded" due to communication problems related to language proficiency that become apparent in the course of initial training. This is because there is no standardized admission process to screen for language proficiency for flight training. Some students make it through initial flight training but begin to struggle in more advanced courses or in more complex flight situations. Inadequate language proficiency may become apparent and students may need to stop their flight training in favor of English language training. If this occurs, students' operational knowledge at that point can vary greatly depending on where in the flight training process he or she is when the interruption occurs. Students in flight

schools may be grounded at any stage in the flight training if language deficiencies become apparent; students have been grounded as far into flight training as the Commercial Multi-Engine course, meaning they have passed at least two FAA check rides with an English Proficient endorsement, and yet were later determined to need additional language training.

After being enrolled, these students were all assessed using an oral proficiency assessment and an adapted ICAO rating scale, with both a language and an operational expert. Test results included learners who demonstrated English language proficiency from high ICAO Pre-Operational Level 3 to low ICAO Elementary Level 2. This represents the typical language ability range of such students who will benefit from an English for Flight Training course.

9.3.1. Goals and objectives

The primary objective of English for Flight Training is to support the language acquisition of NNES flight students and to prevent inadequate language proficiency from interfering or interrupting their flight training. Although many students need more than 120 hours of English instruction, this course intends to improve English skills which are relevant to flight training and communication skills in flight. It does not intend to take the place of ground school or instruction from a CFI.

Conceptual underpinnings: Content Based Instruction (CBI). Since this course is not intended to replace ground school but, rather, to prepare students for it, it should be regarded more as a CBI than a CLIL (Content and Language Integrated Learning) course. Although content acquisition will undoubtedly occur during the course and that is undeniably useful, it is not the only opportunity students will have to learn this information. Exposure to this content, particularly vocabulary, gives students a head-start in ground school and may even give them a confidence boost since they will already be familiar with the topics and language they will encounter. Although ground school presumes no previous knowledge of aviation, most individuals who choose a pilot profession have some foundational knowledge of aircraft, airports, and aerodynamics from their prior general interest in and experience with the subject. English for Flight Training helps to equalize, to some degree, disparity in content and language proficiency that may exist among students. A theme-based CBI approach was used in the development of this course, following quite closely the sequencing of Embry-Riddle's offerings which are freely available online MOOCs: Aviation 101 (www.aviation101.org), discussed below.

Syllabus design: Choosing the content. The Aviation 101 website states, "This is a great opportunity for any high school student who is trying to choose a career and wants to learn more about aviation. It is also ideal for anyone who wants to explore the essentials of aviation, or for experienced pilots to review some of the basics." This description is not far from that of English for Flight's target audience. Aviation 101 includes nine modules, of which aircraft systems, aerodynamics, airports, radio communication and ATC, and aviation weather were used as primary content sources for this course.

Phraseology is not included in the syllabus, and it is recommended to be offered as a separate course like English for VFR (see below), ideally as part of the same training program. However, because very often in radiotelephony communications phraseology is combined with general English (Kukovec, 2008), efforts are made throughout the course to integrate utterances which include phraseology as functional classroom language. For example, when making requests, checking understanding, clarifying, affirming, etc., students can use standardized phraseology lexical units such as *affirmative, read back correct, say again, verify, confirm,* etc.

9.3.2. Activity types

Thematic instruction. One of the first themes in the course is aircraft systems, which covers basic parts of an aircraft, landing gear, flight controls, the aircraft engine, and fuel and oil systems. In each component of the theme, lessons are designed to focus on the six ICAO criteria and elements of basic EAP to prepare students for the classroom environment of ground school.

Lesson particulars: The fuel system. After completing a unit on the engine, the fuel system unit begins with pre-listening activities for brainstorming and vocabulary review. Students are provided with a list of vocabulary words and hear a recorded sentence from the Aviation 101 video which contains the vocabulary word. Using this context and any prior knowledge, groups brainstorm definitions and predict the application or function of the lexical item within the context of the fuel system. The video is accompanied by a listening guide. Typically, videos are viewed up to three times using targeted listening and pausing for review when necessary. Follow-up discussion and activities after the video check comprehension and focus on the most important concepts contained therein. A brief pronunciation exercise follows, emphasizing the uniqueness of /ˈfjuːəl/, as well as a series of exercises drawing awareness to the passive structures heard frequently in the video, for example, "the fuel system *is designed* to bring fuel," "high-wing airplanes *are equipped* with some sort of fuel pump," "gas *is pumped* into a tank." These pronunciation and grammatical structure characteristics are then emphasized in communicative activities for review and recycling of information to increase and ensure understanding.

Each unit concludes with a major culminating activity, usually a form of research, presentation, performance, or creation. In the fuel system unit, for example, students are challenged to provide the narrative voice to a pre-made Pecha Kucha presentation. Students practice alone with printed images of the slides and are then partnered to participate in a speed dating activity. As the master Pecha Kucha presentation plays, students alternate providing the quick twenty-second description of the stage of the fuel system present on the screen. Final presentations are done with partners for the class after significant practice rounds during speed dating. The fast pace and information-dense nature of Pecha Kucha is particularly effective for practicing and developing fluency.

Vocabulary acquisition activities. Vocabulary-focused activities are among the most common in the class, including many vocabulary learning games which force students to paraphrase. For example, flash cards are made for all major vocabulary words throughout the course, and are used in activities in which students describe the vocabulary item to classmates, identify key, related vocabulary which become *taboo* when describing, and a variety of identification games using pictures, blindfolds, and charades. All of these games provide practice in paraphrasing, which is critical to deal with vocabulary comprehension issues when speaking. In fact, ICAO Operational Level 4 description of Vocabulary in the rating scale states that those with Operational Level 4 proficiency "can often paraphrase successfully when lacking vocabulary in unusual or unexpected circumstances" (ICAO Document 9835, 2010). Creative design of materials and activities can create this simulated lack of accessible vocabulary, thereby prompting and requiring students to develop and utilize the skills to resolve this challenge, such as circumlocution and approximation, as well as paraphrasing, common strategies used to compensate for lack for vocabulary during communication (Paramasivam, 2013). Guessing games are particularly effective as they are repetitive, spontaneous, and unpredictable, containing two-way (minimally) interaction in real time, focusing on the outcome instead of the precise language being used to accomplish the outcome (Thornbury, 2008, as cited in Paramasivam, 2013).

Numbers. Accurate communication and comprehension of numbers are particularly difficult challenges in aviation communication due to irregularities in pronunciation and the way numbers are read, despite the standards set forth by ICAO (e.g., *666* should be read as "*six, six, six*," but is often heard on the radio as "*triple six*"). Pronunciation of numbers may be heavily influenced by the way numbers are grouped in a speaker's first language. For example, Sullivan and Girginer (2002) found instances of Turkish speakers saying *1931* as *"one nine thirty one"* instead of *"one, nine, three, one"* reflecting the conventional practice in Turkish. Numbers are extremely prevalent in RTF and other aviation communications, so a thorough understanding of contexts which include numbers and intelligible pronunciation consistent with standards are critically important for pilots. Numbers are covered during the first few days of English for Flight Training, as they will necessarily come up throughout the course and are a useful topic to continuously review and practice.

Lesson particulars: Numbers in aviation. Numbers are introduced through a brainstorming activity and reviewed using a competitive game challenging students to think of all possibilities of number usage. A handout is given which reviews twenty of the most common uses of numbers, e.g., call signs, headings, runways, altimeter settings, including the number, its reference, transmission, and pronunciation. For example, *FL 150* refers to a flight level, is transmitted as *flight level one five zero*, and is pronounced as *flight level "wun fife zee-row."* Students work together to complete a chart, then participate in a gap-fill activity to review. Finally, numbers in aviation are printed and pasted on notecards. The class is divided into two teams, with one representative going to the board. The other teammate describes the word without using it (e.g., *altitude, squawk code, airspeed*), to their representative who must write the correct word and an example of the number in context, e.g., *airspeed, 160 knots*, on the board. The lesson on numbers contains many communicative activities using numbers, and sets the stage for further work with numbers in lessons on ATIS and RTF role plays.

9.3.3. Results and observations

At the end of this course, all students are eligible for a certificate of completion. An important caveat to this certificate is that it represents the completion of the training course and is not a licensing certificate. ICAO (2009) states that trainers should never be involved in the licensing of their own students, as this can potentially compromise the validity of the official high-stakes examination with serious safety implications. It is recommended that certificates are marked to indicate as such. At the bottom of the certificates used in this course, a message is printed which states, "This is a record of completion of training only and is not a certification of English proficiency."

9.4. English for Visual Flight Rules course

English for Visual Flight Rules (VFR) is a course devoted to the development of a more specific register of English in aviation. Communicating over the radio is frequently reported as the most difficult aspect of flight training by NNES. The high cognitive workload and fast-paced chatter can be difficult even for the most experienced pilots. This course provides an opportunity to practice RTF in a low-stakes classroom environment where questions and interruptions provide learning opportunities rather than threats to safety. English for VFR Flight is designed to cover just twenty-five hours, spanning three to four weeks, depending on availability of the students. Ideally, it is offered in conjunction with other aviation English language training courses.

9.4.1. Goals and objectives

The course seeks to develop the necessary language skills to effectively participate in radio communications during VFR Flight, including interacting with ATC with fluency, clarity, and accuracy. Throughout the course, students work on comprehending ATIS broadcasts, responding appropriately to various input in radio communications dialogues, and accurately participating in flight scenarios. To do so, students must be able to communicate in fast-paced, dynamic environments without using nonverbal strategies and among other aural and physical distractions. An additional goal of this course is to provide basic foundational knowledge of flight operations, including the typical phases of a flight and the standard language which is part of each stage. Both NNES and NES have been observed to benefit from this course, taken either before or during the initial stages of flight training.

Conceptual underpinnings. As illustrated throughout this book, aviation English contains a broad range of contexts and applications. While the scope of aviation English extends beyond radiotelephony and phraseology, students can still benefit from focused instruction in this small, highly specialized register. In fact, many programs around the world offer separate "Aviation English" and "Aviation English Terminology," "Aviation English Phraseology" or "Radiotelephony" courses (Aiguo, 2008; Kukovec, 2008). English for VFR Flight is notably a much shorter course than others due to its narrower focus on phraseology for a specific type of flight situation. English for VFR was developed primarily using a Task-Based Language Teaching framework, corresponding to the required communication tasks of a routine VFR Flight. In addition, the course strategically utilizes the skills and knowledge of two subject matter experts.

Dual subject matter experts. As outlined in Chapter 7, according to ICAO Document 9835 (2010), the best qualifications for an aviation language trainer, administrator, or materials developer include a master's degree in language teaching, three years or more of aviation language training, and pilot or controller experience. Particularly in a course such as English for VFR

Flight where producing prescribed phraseology accurately is a primary goal, it was necessary to implement the course with a collaboration between two subject matter experts (SMEs), a qualified language specialist, and an aviation specialist, during which the language specialist provided linguistic support to the operational aviation content. ICAO (2009) states that phraseology should be taught as a distinct discipline in the operational domain by specialized and operationally qualified trainers. An aviation English instructor can support this training by aiding in the development of fluency and pronunciation skills, for example, in the limited register of phraseology. In the case of English for VFR Flight at Embry-Riddle, an English language instructor from an applied linguistics and ESL background with aviation familiarity, and a CFI with an aeronautical science background worked together to present the necessary content knowledge and pedagogical skills to create and implement this course.

To ensure the successful implementation of English for VFR Flight, each instructor needs to serve many roles including the traditional language teacher, the traditional flight instructor, the assessor, the organizer, and the materials developer. Both instructors fulfilled each role to some degree, based on their own expertise in language instruction or flight instruction, and spent many hours collaborating within each role to build and successfully implement the course.

9.4.2. The role of the aviation English instructor

The traditional language teacher's role. Particularly toward the beginning, the instructor must take on more traditional language teaching roles such as enforcing an English-dominant classroom and creating groups/pairs for practice activities as appropriate. Language classrooms are by nature more interactive than traditional lecture-based classrooms; therefore, the language teacher must encourage communication and facilitate a certain level of comfort and confidence involving risk-taking to promote purposeful communication in the classroom. Additionally, the experienced language teacher intuitively knows when it's advantageous to write items on the board, when to ask additional, follow-up questions to verify comprehension, and when to seek elaboration for clarity.

The assessor's role. Through comprehension-based assessments, students' progress is monitored and recorded by the language teacher. During class, students are informally assessed based on their ability to perform tasks and their reliance on outside assistance such as the course book or the instructors. This type of assessment dictates the pace of instruction and movement between units. Additionally, the language teacher and flight instructor work together to perform an assessment of oral proficiency both before and after the course to determine progress, with the language teacher assessing communicative ability and the flight instructor assessing operational accuracy.

Organizer's role. The language teacher is responsible for creating day-by-day lesson plans and sequencing the activities of individual classes to follow the CLT framework and ensure ample opportunities for comprehension, interaction, and, most importantly, production of the target language scenarios.

Material developer's role. Utilizing role-play dialogues written by the flight instructor, the language teacher creates learning activities which allow students to practice the information in a meaningful way. From the operational content provided by the flight instructor in the course book and delivered in class, the language teacher can create games, role-play simulations, and interactive dialogues to practice communications and reinforce content knowledge.

9.4.3. The role of the Certified Flight Instructor (CFI)

The traditional flight instructor's role. As is the case during normal flight instruction responsibilities, the CFI explains technical terminology, specifications of the local airport in use, and the meanings of the standard phraseology used during different phases of VFR Flight. He is the primary source of content knowledge and the primary responder to student questions involving the content. A flight instructor assumes the role of mentor in traditional flight instruction (FAA, 2008), and, therefore, replicates this role also in the English classroom by providing explanations based on actual experience. The CFI is instrumental in guaranteeing the accuracy of the course and providing the students with assurance that what they are learning in the classroom will be relevant in real-world contexts as well.

The assessor's role. The CFI is able to assist the language teacher with assessing students by checking for accuracy. Typically, during the course, the CFI debriefed answers with students and could discern if students were managing to correctly report information they had heard. Additionally, the language teacher and flight instructor work together to perform an assessment of oral proficiency both before, during, and after the course to determine progress, with the language teacher assessing communicative ability and the flight instructor assessing operational accuracy.

Organizer's role. The flight instructor is also instrumental in providing the logistical knowledge of activities during such activities as Observer Flights (see explanation of Observer Flights below), advising the language teacher as to considerations involving the pacing of activities during the flight, and the reasonableness of what is being asked of students. Also instrumental in the organization of the course, the flight instructor informs the sequencing of the syllabus to match, as closely as possible, a full-flight scenario.

Material developer's role. One of the most important roles of the flight instructor is providing assistance during the design or course materials to ensure accuracy. In English for VFR Flight, the CFI created all role-plays, including multiple versions of each phase of flight, following the guidelines set forth in the Flight Communication Standards manual utilized by the university's flight department. Additionally, a glossary was created by the CFI to explain difficult terminology using definitions which were then simplified by the language teacher.

9.4.4. Communicative Language Teaching

Despite the very prescriptive nature of the phraseology and the limited tolerance for deviations from standards, the teaching of phraseology can still follow a communicative, collaborative approach. Within the CLT context and methodology, the CFI can provide information and explanations during the presentation stage. Although the CFI is knowledgeable of and accustomed to explaining this information, the aviation English instructor may need to occasionally intervene with comprehension checks or prompting questions to elicit further explanation or clarification if students appear to be struggling. In general, the amount of time that a CFI spends explaining the content to a group of pre-*ab-initio* NNES pilots may be double or even triple that required for students already enrolled in the flight program at the university. It is important to note that this course ideally serves as a precursor to flight training, so this information will be taught again once students have matriculated.

After the presentation of the phraseology content, with explanations of any new terminology and the actions taken by a plane during specific phases of flight, the instructors are then able to facilitate the practice stage together by circulating among the students and reviewing concepts as necessary. During this stage, students work together with appropriate intervention and corrections provided by the instructors. The resulting student performance is then checked for technical completeness and accuracy in content production by the CFI and for accuracy in language production by the aviation English instructor. The skillsets and expertise of both instructors are utilized effectively in this way, and students are provided with adequate content and technical skills as well as language support throughout the lesson.

Syllabus design: Choosing the content. The syllabus of this course is designed around a typical flight scenario and the acquisition of the knowledge and skills necessary to complete it correctly and safely. The course begins with several classes introducing students to ICAO phonetic alphabet, followed by an explanation of ATIS broadcasts. ATIS broadcasts contain necessary information specific to an airport such as weather conditions and active runways. Following a few intensive sessions on ATIS, the course then follows the phases of a typical full-flight: Ground Operations, Departure Operations, Practice Area Operations, and Arrival Operations. By following the full-flight chronologically, a cohesive learning experience is created for students which mimics, as closely as possible, real-world situations.

Designing the tasks. Target tasks related to flight training which are practiced in English for VFR Flight are based on the fundamentals of radio communication, namely the need to listen to and apply ATIS broadcasts, and to comprehend and respond appropriately to instructions from ATC. Additionally, communication between pilots during times such as Practice Area Operations and flight student-flight instructor communications in the flight deck are also addressed. Pedagogic strategies to accomplish achieving these target tasks include a combination of listening comprehension, role-play, and simulation activities. As students engage in these various learning activities, time allotments and constraints are managed as closely as possible to approximate the precision that is necessary on the flight deck, carefully timed to avoid overlapping talk and strictly organized to follow a prescribed sequence of communication (Nevile, 2007).

Instructional materials. The primary course text used was created by the instructors. This text serves mainly as a reference rather than a traditional workbook with questions and activities. Included within the course book are role-play dialogues to provide example scenarios of flight operations, ICAO phonetic alphabet, an airport diagram of the local airport in use, a local practice area diagram, a glossary of terms, blank tables to fill in information from ATIS broadcasts, and handouts for guided observer flights.

To simulate the real-world scenarios that students currently experience or will experience once they enter flight training, maps of the local airport were enlarged and printed on sturdy material to cover the tables in the classroom. Students use models of Cessna 172s together with the enlarged airport maps as they participate in role-plays not only verbally, but visually as well. In this way, familiarity with the layout of the local airport, including runway names and locations, traffic patterns, and landmarks, is also accomplished and reinforced.

9.4.5. Activity types

Automatic Terminal Information Service. Mastering the ability to comprehend and apply an ATIS broadcast is a challenging and often time-consuming task, even for native speakers of English.

Lesson particulars. To facilitate this process for the students of English for VFR Flight, the components of an ATIS were broken down into three categories of importance. Each category is thoroughly explained by the CFI, but listening activities are systematically varied in levels of difficulty by asking students to initially listen only for certain components. The categories, e.g., sky conditions, are considered as a variable slot with a specified number of possible answers, e.g., *few, scattered, broken, clear*. In addition to the meanings of each possible answer, students are taught shorthand techniques for recording this information in an ATIS table prepared by the language instructor. Students are encouraged to deviate from the table and shorthand as they discover what works best for their own comprehension and retention of information. Using an airport of personal relevance, students conduct research and create their own ATIS broadcast which is played for the class as a listening comprehension exercise. Each day, new components of the ATIS broadcast are examined and practiced in the same manner, until finally, students are able to gather all the necessary information from the broadcast consistently and without error. In a recently conducted English for VFR Flight class, three one-hour class periods were devoted solely to ATIS, then the current ATIS was listened to and reviewed daily for the remainder of the course. However, by the end of just the second week, students could easily and accurately comprehend and report all information from the ATIS.

Radio Communication during a flight. Once ATIS reports have been mastered, the content of the lessons become organized based on four distinct segments (or "legs") of a flight: Ground, Departure, Practice Area, and Arrival. The methodology follows the basic CLT framework of Present, Practice, and Produce.

Lesson particulars. To begin, the instructors and the students dissect the example role-play given in the course book, with explanations of content from the CFI and awareness of language patterns noted by the Aviation English instructor. Importantly, those features within the dialogue which might vary are pointed out and other alternative options are provided by the instructors. Depending on the content being presented, the CFI provides relevant information. For example, before discussing the radio communications associated with Arrival Operations, an explanation of airport traffic patterns specific to the local airport is given. Students are provided with exact transcripts of the dialogue to establish awareness-raising conditions (Paramasivam, 2013). In this awareness-raising stage, attention, noticing, and understanding of details and patterns are the focus.

The next stage is the appropriation stage during which learners need to begin demonstrating their own ability to take control of the language (Paramasivam, 2013). Students practice the role-play dialogue with a partner, almost to the point of memorization. Those students who are more advanced and comfortable with options can begin substituting other possibilities into the variable slots. Finally, after substantial rehearsal of the dialogue, one partner at a time is challenged to close his/her book and reproduce the dialogue again. Partners switch back and forth until they are comfortable producing the role-play without any reference to the book. At various times, the instructor may ask students to switch partners or practice in front of the class if more focus or reinforcement seems to be needed or partners are not working well together.

In the future, to increase the validity of the role-plays, recordings of actual flight deck communications between flight instructors and students will be collected, analyzed, and made available for classroom activities and follow-up studies. Tuccio et al.'s (2016) approach using Collaborative Transcription and Repair-Based Learning (CTRBL) will be utilized to allow pilot-learners to transcribe audio recordings and then attempt to repair problems, e.g., communication errors or deficiencies by marking corrections and revisions on the transcripts. Listening to actual recordings increases the likelihood of authenticity in the students' role-plays as observed and recommended by Stokoe (2014).

For pilots, the capacity to manage a high workload is necessary to avoid work overload and to ensure that operations are successfully and safely accomplished. Molesworth and Estival (2015) found that NNES' performance deteriorates when radio communication is combined with other tasks which create a heavy workload. By gradually increasing workload during a classroom lesson, instructors can help students develop the capacity to manage heavy workload demands (FAA, 2008). Additionally, pilots and controllers must process large amounts of visual information while performing other linguistic tasks such as communicating with crew members, other aircraft in the sky, and monitoring instruments (Uplinger, 1997). Therefore, once all stages of a flight have been presented, practiced, and produced by students, aural and physical distractions are added to the scenarios to more accurately simulate the real-world environment and workload realities of flying an aircraft.

Pilots are not able to rely on nonverbal cues such as facial expressions or gestures during radio communications (ICAO, 2010). Additionally, the flight deck of a Cessna 172 can be quite loud and, despite the multitude of things happening around them, pilots must maintain strong situational awareness of and focus upon their immediate environment, while, at the same time, maintaining peripheral awareness of relevant, developing events. ICAO (2010) recommends "the need for training and testing to provide voice-only settings to exercise or demonstrate language proficiency as well as face-to-face settings that allow broader uses of language" (4.6). Creating this training scenario and replicating these particular realities of RTF in a classroom are not easy, but by utilizing such creative teaching strategies as slowly adding distractions by playing the broadcast from LiveATC.net of the local ATC tower, TRACON; having students stand back-to-back to prevent eye contact; and asking those students acting as pilots during the role-play to toss a small, inflatable basketball to other pilots around the room, the instructors were able to successfully create a noisy, busy, and dynamic—a realistic!—"flight deck" within the classroom.

Additionally, in order to simulate radio communication more accurately, walkie-talkies are utilized during role-plays. Students can physically separate by being in different rooms, ensuring that they are not able to rely on any visual cues while communicating. Communication between pilots and controllers lacks the paralinguistic features of face-to-face communication, so tools like body language and visual cues are not facilitating comprehension during communication. While using walkie-talkies, students must, instead of these visual cues, rely more on paraphrasing and clarification techniques to communicate effectively (Bullock, 2015). The use of frequencies, the sound of radio static, and the complications of communicating absent physical proximity creates an even greater sense of authenticity. At this point, learners are now in the autonomy stage of using the language. They are able to perform with minimal assistance, i.e., without recourse to the textbook or other support, and respond to unpredictable and spontaneous utterances from the instructors or classmates (Paramasivam, 2013).

Full-flight scenarios. The culminating pedagogical experience in the course is a series of role-plays involving all students participating in various roles of a full-flight scenario. A large-scale map of the local airport is placed on top of a table in the center of the room and each student is assigned an ATC role of either Clearance, Ground, Tower, Departure, or Approach. Students initially group and work together with others assigned the same ATC role to review and familiarize themselves with that controller's role during a full-flight scenario. One student from each role then wears a name tag indicating they are currently acting as Clearance, Tower, etc., and all other students take a small replica of a Cessna 172, choose a call sign and label it on their plane, and prepare to play the role of pilots. From there, students playing the role of pilots take turns asking for clearance from the student playing the role of Clearance Delivery Controller, then communicating with the Tower Controller while another pilot is now asking for clearance. The sequence continues until eventually all students are interacting at the same time, needing to maintain awareness of what's happening in and around the airfield, which controllers are currently available for communication, and what the appropriate sequence of communication is. Throughout the simulation activity, the instructors assume the role of careful and alert monitors to maintain the flow of the role-play and to ensure accuracy of the performance of all roles. Once all pilots have asked for clearance, received taxi instructions, taxied to the runway, taken off, communicated with other pilots in the practice area, contacted approach to return to the airport, landed, and taxied back to the ramp, the ATC roles are handed off to another student group and the exercise is repeated. Again, the classroom is transformed into a loud, hectic, and fast-paced environment simulating as closely as classroom-constraints permit, a full-flight scenario.

Observer flights. Throughout the course, students participate in four observer flights to experience hands-on flight operations and to become familiar with the local flight facilities. Although each observation flight will be full-flight, students are required to focus on different phases and aspects of the flight each time to reinforce the particular skills learned in the classroom. For example, Observation Flight #1 focuses on Ground and Departure Operations and asks the student to notice and record information on ATIS, clearance delivery, taxi instructions, takeoff directions, etc. (See Table 9.1.) By the time they take the observer flight, students have, of course, been presented with the significance and details of each aspect of these operations and have practiced them in a classroom setting. The observer flight affords the opportunity to experience and apply the classroom training in the real-life context.

Table 9.1 Observer Flight #1 handout from course book [Observer Flight #1 is one of four Observer Flights students take during the course.]

Observer Flight #1

Ground and Departure Operations

This observer flight will help you in the development of proper ground and departure communication techniques. Pay close attention to the communications between the pilots and Air Traffic Control (ATC). If there are questions you are unable to answer, feel free to ask the pilot **after the flight is over. Do not ask any questions while the pilot is flying.**

You will need the following handouts: **(1) Observer Flight #1, 2) KDAB Airport Diagram 3) ATIS Tables ATIS and Clearance**

1. Complete the ATIS handout for both departure and arrival.
2. Squawk code: _____
3. Clearance altitude: _____
4. North or South practice area. (circle one)

Taxi and Takeoff Phases

1. Highlight the taxi instructions on the airport diagram.
2. Were there any **hold short** instructions and on what intersection?
3. Which runway did you takeoff from? Circle the runway.
4. After departure, where did the airplane fly? For example, shoreline southbound, shoreline northbound, etc.

9.5. Concluding remarks

At flight training institutions, if students are assessed and found to lack the appropriate prerequisite language skills or the proficiency level necessary to begin flight training, additional, supplemental training experience should be recommended which can include courses such as the three described in this chapter, Aviation Topics, English for Flight Training, and English for VFR Flight. It is understood that one course alone might not provide adequate supplemental learning for these students or be a complete solution to their language training needs to enable them to achieve the requisite skill level to continue with their flight training. It is to the responsibility of flight schools to develop the type of program that works best for their student body. Certainly, as English continues to be the *lingua franca* of flight training, more careful attention needs to be given to this issue and the many safety implications present. The development of relevant, comprehensive, and effective training strategies to ensure competency of aviation personnel in their various roles must be an important priority.

Notes

Chapter 2

1. International Federation of Airline Pilots (IFALPA); International Air Transport Association (IATA) International Federation of Air Traffic Controllers Association (IFATCA) and the European Organization of Air Navigation (Eurocontrol).
2. ICAO Guidance manual, Document 9835, *Manual on the Implementation of ICAO Language Proficiency Requirements,* contains ICAO language SARPS, and provides support aimed at the aviation community, ICAO Member States, airlines, air navigation service providers, and training programs that must implement the SARPS into their national regulations and programs.
3. As an example of why this amendment was important we can consider ICAO phrase, "Line up and wait," an air traffic control instruction to a pilot maneuvering an aircraft into position prior to receiving clearance to take-off. Until 2010, the US Federal Aviation Administration terminology for the same ground action was, "Taxi into Position and Hold" (FAA Notice N JO 7210.754 September 20, 2010). Two different phrases for the same ground action were confusing and potentially dangerous if international pilots flying into and out of US airports were not familiar with the FAA phraseology. The amendments adopted in 2003 specified that ICAO phraseology, found in ICAO Document 4444—Procedures for Air Navigation Services, Air Traffic Management and Annex 10, Vol. II, is required.
4. *ICAO Rating Scale* is published in Appendix 1 to Annex I—*Personnel Licensing.* The original SARPS can also be found in Appendix 1 of ICAO Document 9835 Manual, along with all of the Standards and Recommended Practices related to language, consolidated in Document 9835 for ease of reference.
5. *ICAO Annex 6 Part III Chapter 1 paragraph 1.1.3* states: "Operators shall ensure that flight crew members demonstrate the ability to speak and understand the language used for radiotelephony communications as specified in Annex 1."
6. Features of ICAO Rating Scale are addressed in Section 4.5 of the second edition of ICAO Document 9835 Manual.

Chapter 7

1. The FAA addresses some of these issues in FAA Advisory Circular AC 60-28B: https://www.faa.gov/documentLibrary/media/Advisory_Circular/AC_60-28B.pdf
2. See FAA Advisory Circular AC 60-28B: https://www.faa.gov/documentLibrary/media/Advisory_Circular/AC_60-28B.pdf
3. According to FAA, "in the case of an applicant for a Flight Engineer, Flight Navigator, Air Traffic Control Tower Operator, Mechanic, Repairman, Parachute Rigger, or Remote Pilot Certificate who does not meet this requirement, the Administrator *may place any limitation on the certificate that he or she considers necessary for safety.*" (emphasis added, http://fsims.faa.gov/PICDetail.aspx?docId=8900.1,Vol. 5,Ch. 14,Sec. 1)

Chapter 8

1. As aviation handles security issues and is necessarily private, some procedures are prohibited to those outside the workplace.
2. Available at: https://moodle.org/
3. Available at: http://exelearning.net/descargas/
4. Available at: https://www.blackboard.com/index.html
5. Available at: https://www.adobe.com/training/elearning.html
6. Available at: https://www.braincert.com/online-virtual-classroom
7. Available at: https://www.learncube.com/virtual-classroom-software.html
8. Available at: https://www.wiziq.com/virtual-classroom/
9. Available at: https://zoom.us/
10. Airport charts are publicly available at websites such as AOPA Airports, AirNav.com, and SkyVector.com.

References

"Accident of the aircraft 5B-DBY of Helios Airways." (2005). Flight HCY522, BOEING 737-31S, on August 14, 2005. Accident Investigation Report 11/2006. Available at http://www.aaiasb.gr/imagies/stories/documents/11_2006_EN.pdf

Aiguo, W. (2008). Reassessing the position of aviation English: From a special language to English for specific purposes. *Iberica*, 15, pp. 151–164.

"Air Canada Boeing 767 C-GAUN Accident, Gimli, Manitoba." (1983). Report of the Board of Inquiry. Available at http://data2.collectionscanada.gc.ca/e/e444/e011083519.pdf

"Aircraft Accident Report." (1998). ALPA Engineering and Air Safety Report 26. Available at http://archives.pr.erau.edu/ref/Tenerife-ALPAandFIP.pdf

Albritton, A. (2007). ICAO language proficiency in ab-initio flight training. *Second ICAO Aviation Language Symposium, Montréal*. Available at https://www.icao.int/Meetings/AMC/MA/Second%20ICAO%20Aviation%20Language%20Symposium%20(IALS-2)/24.Albritton.pdf

"American Airlines Flight 965—Aircraft Accident Report." (1995). "Controlled Flight into Terrain. Being 757-223, N651AA; December 20. Aeronautical Civil of the Republic of Colombia, Santa Fe de Bogota, D.C. Colombia." Available at https://skybrary.aero/bookshelf/books/1056.pdf

Anthony, L. (2018). AntConc (Version 3.4.3) [Computer Software]. Tokyo: Waseda University. Accessed July 9. Available at http://www.laurenceanthony.net/

Arinik, H. (2015). Technology as a learning aid. In *International Civil Aviation English Association Annual Conference*, Santa Maria Island, The Azores.

(ARSA) Aeronautical Repair Station Association. (2017). Speaking the FAA's language – guidance on English proficiency. Available at http://arsa.org/english-proficiency/

Asención-Delaney, Y., & Collentine, J. (2011). A multidimensional analysis of a written L2 Spanish corpus. *Applied Linguistics*, 32(3), 299–322.

"Assembly Resolution A32-11: Establishment of an ICAO universal safety oversight audit programme." (1998). Montreal: ICAO.

"Assembly Resolution A32-16: Proficiency in the English language for radiotelephony communications." (1998). Montreal: ICAO.

"Avianca 052: Aircraft Accident Report. January 25, 1990." (1990). NTSB/AAR-91/04. Washington, DC: National Transportation Safety Board.

"Aviation Occurrence Categories, Definitions, and Usage Notes." (2011). ICAO CAST Document. July. International Civil Aviation Organization. Montreal.

Baker, P. (2006). *Using corpora in discourse analysis*. New York: Continuum.

Barbieri, F. (2008). Patterns of age-based linguistic variation in American English. *Journal of sociolinguistics*, 12(1), 58–88.

Barshi, I. (1997). Effects of linguistic properties and message length on misunderstandings in aviation communication. Unpublished PhD dissertation, University of Colorado.

Barshi, I., & Healy, A. (1998). Misunderstandings in voice communication: Effects of fluency in a second language. In A. F. Healey & L. E. Bourne (Eds.), *Foreign language learning: Psycholinguistic studies in training and retention* (pp. 161–192). Mahwah, NJ: Erlbaum.

Benjamin, R. (2011). Reconstructing readability: Recent developments and recommendations in the analysis of text difficulty. *Educational Psychology Review*, 24(1), 63–88. Available at http://dx.doi.org/10.1007/s10648-011-9181-8

Berber Sardinha, T., & Veirano Pinto, M. (Eds.) (2014). *Multi-dimensional analysis, 25 years on*. Amsterdam: John Benjamins Publishing Company.

Berber Sardinha, T., & Veirano Pinto, M. (Eds.) (2019). *Multi-dimensional analysis: Research methods and current issues*. London: Bloomsbury.

Berber Sardinha, T., Kauffmann, C., & Acuzo, C. (2014). A multi-dimensional analysis of register variation in Brazilian Portuguese. *Corpora*, 9(2), 239–271.

Bhatia, V. K. (2008). Genre analysis, ESP and professional practice. *English for Specific Purposes*, 27, 161–174.

Biber, D. (1988). *Variation across speech and writing*. Cambridge: Cambridge University Press.

Biber, D. (1993). Representativeness in corpus design. *Literary and Linguistic Computing*, 8(4), 243–257.

Biber, D. (1995). *Dimensions of register variation: A cross-linguistic perspective*. Cambridge: Cambridge University Press.

Biber, D. (2006). *University language: A corpus-based study of spoken and written registers*. Amsterdam: John Benjamins.

Biber, D., & Burgess, J. (2001). Historical shifts in the language of women and men. In S. Conrad & D. Biber (Eds.), *Variation in English: Multi-dimensional studies* (pp. 21–37). London: Longman.

Biber, D., & Conrad, S. (2009). *Register, genre, and style*. Cambridge: Cambridge University Press.

Biber, D., Conrad, S., & Cortes, V. (2004). *If you look at...*: Lexical bundles in university teaching and textbooks. *Applied Linguistics*, 25(3), 371–405.

Biber, D., Conrad, S., & Reppen, R. (1998). *Corpus linguistics: Investigating language structure and use*. Cambridge: Cambridge University Press.

Biber, D., Davies, M., Jones, J. K., & Tracy-Ventura, N. (2006). Spoken and written variation in Spanish: A multi-dimensional analysis. *Corpora*, 1, 7–38.

Biber, D., & Hared, M. (1992). Literacy in Somali: Linguistic consequences. *Language Variation and Change*, 4, 41–75.

Biber, D., Johansson, S., Leech, G., Conrad, S., & Finnegan, E. (1999). *Longman grammar of spoken and written English*. Harlow: Pearson.

Biber, D., Reppen, R., & Friginal, E. (2010). Research in corpus linguistics. In R. B. Kaplan (Ed.), *The Oxford handbook of applied linguistics* (2nd ed.) (pp. 548–570). Oxford: Oxford University Press.

Bieswanger, M. (2016). Aviation English: Two distinct specialised registers? In C. Schubert & C. Sanchez-Stockhammer (Eds.), *Variational text linguistics: Revisiting register in English* (pp. 67–85). Berlin: DeGruyter.

Billings, C. E., & Cheaney, E. S. (1981). Information transfer problems in the aviation system. NASA Technical Paper 1875. Washington, DC: National Aeronautics and Space Administration.

REFERENCES

"Boeing report and technical outlook." (2018). *2018 Boeing pilot and technician outlook. Boeing*. Available at https://www.boeing.com/commercial/market/pilot-technician-outlook/

Borowska, A. P. (2017). *Avialinguistics: The study of language for aviation purposes*. Frankfurt am Main; New York: Peter Lang Edition.

Bowles, H. (2014). How about getting those guys in the tower to speak English? Miscommunication, ELF and Aviation Safety. *Textus*, 1, 85–100. Available at http://www.rivisteweb.it/doi/10.7370/77487

Brown, D. (2007). *Teaching by principles: An interactive approach to language pedagogy*. New York: Pearson.

Brown, H., & Bradford, A. (2016). EMI, CLIL, & CBI: Differing approaches and goals. *Japan Association for Language Teaching, Winc Aichi, Nagoya, Japan*. Available at http://jalt-publications.org/sites/default/files/pdf-article/jalt2016-pcp-042.pdf

Brown, H., & Lee, H. (2015). *Teaching by principles: An interactive approach to language pedagogy* (4th ed.). London: Pearson Education ESL.

Bullock, N. (2015). Defining meaningful material for the teaching of English in aeronautical communications. In A. Borowska & E. Adrian (Eds.), *Changing perspectives on aviation English training* (pp. 8–34). Warsaw: University of Warsaw.

Bullock, N. (2017). A re-evaluation of washback for learning and testing language in aeronautical communications. *International Civil Aviation English Association, Embry-Riddle Scholarly Commons*. Available at https://commons.erau.edu/icaea-workshop/2017/monday/19

Bygate, M., Skehan, P., & Swain, M. (Eds.) (2001). *Researching pedagogic tasks: Second language learning, teaching, and testing*. London: Longman.

CAAC. (2017). *Statistical bulletin of civil aviation industry development in 2016*. Beijing: CAAC.

CAAC. (2018). *Annual report of Chinese civil aviation pilot development 2017*. Beijing: CAAC.

"Call Home American English Speech." (2004). University of Pennsylvania. Available at http://www.ldc.upenn.edu/Catalog/catalogEntry.jsp?catalogId=LDC97S42

Campbell-Laird, K. (2006). Pedagogical approaches to aviation phraseology and communication training in collegiate flight programs. *Collegiate Aviation Review*, 24(1), 25–41. Available at https://search-proquest-com.ezproxy.libproxy.db.erau.edu/docview/860951817/fulltextPDF/F53014D99FB04359PQ/1?accountid=27203

Carrell, P. L. (1987). Readability in ESL. *Reading in a Foreign Language*, 4(1), 21–40.

Carter, R., & McCarthy, M. (2017). Spoken grammar: Where are we and where are we going? *Applied Linguistics*, 38(1), 1–20.

Chafe, W. L. (1985). Linguistic differences produced by differences between speaking and writing. In D. R. Olson, N. Torrence, & A. Hildyard (Eds.). *Literature, language, and learning: The nature and consequences of reading and writing* (pp. 105–123). Cambridge: Cambridge University Press.

Cheng, W., Greaves, C., & Warren, M. (2008). *A corpus-driven study of discourse intonation*. Amsterdam: John Benjamins.

"China Airlines Accident Report." (2002). In-Flight Breakup China Airlines Flight C1611. Aviation Occurrence Report, I, ASC-AOR-05-02-001. Republic of China: Aviation Safety Council.

Circelli, D. (2018). National Training Aircraft Symposium tackles pilot shortage and critical aviation industry issues. *Embry-Riddle Newsroom*. Available at https://news.erau.edu/headlines/national-training-aircraft-symposium-tackles-pilot-shortage-and-critical-aviation-industry-issues/

Clark, B. (2017). *Aviation English research project: Data analysis, findings, and best practice recommendations*. Available at https://publicapps.caa.co.uk/docs/33/CAP1375%20Mar17.pdf

Connor-Linton, J. T. (1989). Crosstalk: A multi-feature analysis of Soviet-American spacebridges. Unpublished PhD dissertation, University of Southern California.

Conrad, S., & Biber, D. (Eds.) (2001). *Variation in English: Multi-dimensional studies*. London: Longman.

"Convention on International Civil Aviation." (1944). *International Civil Aviation Organization convention, Chicago, IL. December 1944*. Montreal: ICAO.

Cowie, C., & Murty, L. (2010). Researching and understanding accent shifts in Indian call center agents. In G. Forey & J. Lockwood (Eds.), *Globalization, communication and the workplace* (pp.125–146). London: Continuum.

Crashborn, O. (2008). Open access to sign language corpora. In O. Crashborn, T. Hanke, E. Efthimiou, I. Zwitserlood, & E. Thoutenhoofd (Eds.), *Construction and exploitation of sign language corpora* (pp. 33–38). Third workshop on the representation and processing of sign language. Paris: European Language Resources Association (ELRA).

Croft, J. (2015). Asia-Pacific offers much opportunity for expat pilots. *Aviation Week and Space Technology*. Available at http://aviationweek.com/commercial-aviation/asia-pacific-offers-much-opportunity-expatriate-pilots

Crossley, S., Allen, D., & McNamara, D. (2011). Text readability and intuitive simplification: A comparison of readability formulas. *Reading in a Foreign Language*, 23(1), 84–101. Available at http://dx.doi.org/10.1177/1362168811423456

Crossley, S. A., Greenfield, J., & McNamara, S. D. (2008). Assessing text readability using cognitively based indices. *TESOL Quarterly*, 12(3), 475–512.

Crossley, S. A., & McNamara, D. S. (2009). Computational assessment of lexical differences in L1 and L2 writing. *Journal of Second Language Writing*, 18(2), 119–135.

Cushing, S. (1994). *Fatal words: Communication clashes and aircraft crashes*. Chicago: University of Chicago Press.

Cutting, J. (2012). English for airport ground staff. *English for Specific Purposes*, 31(2), 3–13.

D'Ausilio, R. (1998). *Wake up your call center: How to be a better call center agent*. Indiana: Purdue University Press.

Dahlström, N., Laursen, J., & Bergström, J. (2008). Crew resource management, threat and error management, and assessment of CRM skills: Current situation and development of knowledge, methods and practices. Report for the Swedish Civil Aviation Authority. Lund University School of Aviation.

Daley, S. (May 23, 2000). Pilots just say no to English-only. *New York Times*, A4.

Davies, M. (2018). WordandPhrase.Info [website]. Accessed August 12, 2018. Available at corpus.byu.edu

Dennis, W. (2015). Chinese ATC to adopt English-only policy. AIN Online. Accessed November 26, 2017. Available at AINONLINE.com

"Document 9835." (2004). *Manual on the implementation of ICAO language proficiency requirements* (1st ed.). Montreal: ICAO.

REFERENCES

"Document 9835." (2010). Manual on the Implementation of ICAO language proficiency requirements (2nd ed.). Montreal: ICAO.

Doolan, S. M. (2014). Comparing language use in the writing of developmental generation 1.5, L1, and L2 tertiary students. *Written Communication*, 31(2), 215–247.

Dougan, M. (2016). *A political economy analysis of China's civil aviation industry.* New York: Routledge.

Drury, C. G., Ma, J., & Marin, C. V. (2005). Language error in aviation maintenance, final report for Federal Aviation Administration. Buffalo, NY: University of Buffalo. Available at https://www.faa.gov/about/initiatives/maintenance_hf/library/documents/media/human_factors_maintenance/maint_language_final.pdf

Dumitru, M., & Boşcoianu, M. (2015). Human factors contribution to aviation safety. Available at http://www.afahc.ro/ro/afases/2015/afases_2015/air_force/Dumitru_%20Boscoianu.pdf

Dusenbury, M., & Bjerke, E. (2013). Predictive power of English Testing: Training international flight students. *Journal of Aviation/ Aerospace Education & Research*, 23(1). Available at http://commons.erau.edu/jaaer/vol23/iss1/5

Emery, H. (2015). Aviation English for the next generation. In A. Borowska & E. Adrian (Eds.), *Changing perspectives on aviation English training* (pp. 8–34). Warsaw: University of Warsaw.

Emirates. (2015). Emirates cabin crew now 20,000 strong. *Emirates*. Available at https://www.emirates.com/media-centre/emirates-cabin-crew-now-20000-strong

Engle, M. (2000). Culture in the cockpit – CRM in a multicultural world. *Journal of Air Transportation World Wide*, 5(1), 107–114.

EnglishUSA. (2018). Membership requirements. *EnglishUSA the American Association of Intensive English Programs*. Available at https://www.englishusa.org/page/MembershipRequiremen

Estival, D. (2018). Challenges of aviation communication. Paper presented at Georgia State University's Brown Bag Session for Graduate Students, October 1, 2018. Atlanta, GA.

Estival, D., & Molesworth, B. R. C. (2009). A study of EL2 pilots radio communication in the general aviation environment. *Australian Review of Applied Linguistics*, 32(3), 24.1–24.16.

Estival, D., Farris, C., & Molesworth, B. (2016). *Aviation English: A lingua franca for pilots and air traffic controllers.* New York: Routledge.

"FAA Notice N JO 7210.754." (2010). FAA Line Up and Wait (LUAW) Operations. Notice NJO 7210.754. Available at htps://www.faa.gov/regulations_policies/orders_notices/index.cfm/go/document.information/documentID/319151

"FAA Risk Management Handbook: FAA-H-8083-2." (2009). Federal Aviation Agency: Washington, DC.

Farris, C., & Turner, C. (2015). *Beyond the ICAO language proficiency requirements: A way forward for aviation language testing research.* Montreal: Department of Integrated Studies in Education, McGill University.

Federal Aviation Administration (FAA). (2008). *Aviation instructor's handbook* (Publication No. FAA=H-8083-9A). Washington, DC. Available at https://www.faa.gov/regulations_policies/handbooks_manuals/aviation/aviation_instructors_handbook/media/faa-h-8083-9a.pdf

Federal Aviation Administration (FAA). (2013). *English Language Skill Standards Required by 14 CFR Parts 61, 63, and 65*. (AC No: 60-28A).

Ferrer, R., Empinado, J., Calico, E., & Floro, J. (2017). Standard and nonstandard lexicon in aviation English: A corpus linguistic study. Conference Paper: Pacific Asia Conference on Language, Information and Computation, Tagaytay City, Philippines.

Firth, J. (1957). *Papers in linguistics*. Oxford: Oxford University Press.

Fischer, U., & Orasanu, J. (1999). Say it again, Sam! Effective communication strategies to mitigate pilot error. Proceedings of the 10th International Symposium on Aviation Psychology, Columbus, OH.

Flesch, R. (1948). A new readability yardstick. *Journal of Applied Psychology*, 32, 221–233.

Flight Safety Foundation. (1998). Boeing 757 CFIT accident at Cali, Colombia, becomes focus of lessons learned. *Flight Safety Digest, May–June 1998*. Available at https://flightsafety.org/fsd/fsd_may-june98.pdf

Flin, R., O'Connor; P., & Crichton, M. (2008). *Safety at the sharp end: A guide to non-technical skills*. Surey: Ashgate Publishing Company.

Flowerdew, L. (2005). An integration of corpus-based and genre-based approaches to text analysis in EAP/ESP: Countering criticisms against corpus-based methodologies. *English for Specific Purposes*, 24(3), 321–332.

Friedman, T. L. (2005). *The world is flat*. New York: Picador.

Friginal, E. (2008). Linguistic variation in the discourse of outsourced call centers. *Discourse Studies*, 10(6), 715–736.

Friginal, E. (2009). *The language of outsourced call centers: A corpus-based study of cross-cultural interaction*. Amsterdam: John Benjamins.

Friginal, E. (2013a). Linguistic characteristics of intercultural call center interactions. In G. Nelson & D. Belcher (Eds.), *Critical and corpus-based approaches to intercultural rhetoric* (pp. 127–153). Ann Arbor: University of Michigan Press.

Friginal, E. (2013b). 25 Years of Biber's MDA: Introduction to the special issue and an interview with Douglas Biber. *Corpora*, 8(2), 137–152.

Friginal, E. (2015). Telephone interactions: A multidimensional comparison. In V. Cortes & E. Csomay (Eds.), *Corpus-based research in applied linguistics: Studies in honor of Doug Biber* (pp. 25–48). Amsterdam: John Benjamins Publishing Company.

Friginal, E. (2018). Quantifying cross-cultural communication. Paper presented at the American Association for Applied Linguistics (AAAL) Conference 2018, March 24–27, 2018, Chicago, IL.

Friginal. E., & Hardy, J. A. (2014a). *Corpus-based sociolinguistics: A guide for students*. New York: Routledge.

Friginal, E., & Hardy, J. A. (2014b). Conducting Biber's corpus-based multi-dimensional analysis using SPSS. In T. Berber Sardinha & M. Veirano Pinto (Eds.), *Multi-dimensional analysis, 25 years on* (pp. 297–316). Amsterdam: John Benjamins Publishing Company.

Friginal, E., Lee, J. J., Polat, B., & Roberson, A. (2017). *Exploring spoken English learner language using corpora: Learner talk*. London: Palgrave Macmillan.

Gabrielatos, C., & Sarmiento, S. (2006). Central modals in an aviation corpus: Frequency and distribution. *Letras De Hoje*, 41(2), 215–240.

Garcia, A. (2016). Air traffic communications in routine and emergency contexts: A case study of Flight 1549 'miracle on the Hudson'. *Journal of Pragmatics*, 106, 57–71.

Gladwell, M. (2008). *Outliers: The story of success*. New York: Little, Brown and Company.

Goffman, E. (1967). *Interaction ritual: Essays on face to face behavior*. New York: Doubleday.

Goguen, J., Linde, C., & Murphy, M. (1986). *Crew communications as a factor in aviation accidents*. Washington, DC: NASA Ames Research Center.

Graesser, A. C., & McNamara, D. S. (2011). Computational analyses of multilevel discourse comprehension. *Topics in Cognitive Science*, 3(2), 371–398.

Greenfield, J. (2004). Readability formulas in an EFL context: Are they valid for Japanese speakers? Unpublished doctoral dissertation, Temple University, Philadelphia, PA. (University Microfilms No. 99-38670).

Harris, J. (2012). Aviation's 21st century tower of Babel. *Aerospace International*. Available at www.aerosociety.com

Hawkins, F. H. (1993). *Human factors in flight* (2nd ed.). Surrey: Ashgate Publishing Company.

Hayman J. (2010). Talking about talking: Comparing the approaches of intercultural trainers and language teachers. In G. Forey & J. Lockwood (Eds.), *Globalization, communication and the workplace* (pp.147–158). London: Continuum.

Helmreich, R.L. (1994). Anatomy of an airline accident. *International Journal of Aviation Psychology*, 4(3), 265–284.

Helmreich, R. L., & Foushee, H. C. (1993). Why crew resource management? Empirical and theoretical bases of human factors training in aviation. In E. Wiener, B. Kanki, & R. Helmreich (Eds.), *Cockpit resource management* (pp. 3–45). San Diego, CA: Academic Press.

Helmreich, R. L., & Merritt, A. C. (1996). Cultural issues in Crew Resource Management training. In *ICAO Global Human Factors Seminar*, Auckland.

Helmreich, R. L., Merritt, A. C., & Wilhelm, J. A. (1999). The evolution of crew resource management training in commercial aviation. *International Journal of Aviation Psychology*, 9(1), 19–32.

Henley, I. M. (2003). *Aviation education and training: Adult learning principles and teaching strategies*. Surrey: Ashgate Publishing Company.

Hofbauer, K., Petrik, S., & Hering, H. (2008). The ATCOSIM Corpus of Non-Prompted Clean Air Traffic Control Speech. *LREC*. Available at https://www.spsc.tugraz.at/tools/atcosim

Hofstede, G. (1980). Culture's consequences. International differences in work related values. Accessed October 26, 2012. Available at http://geerthofstede.com/national-culture.html

Hofstede, G. (1991). *Culture's consequences: International differences in work related values*. Beverly Hills, CA: Sage.

Holder, B., Hutchins, E., & Nomura, S. (2002). The ecology of language practices in worldwide airline flight deck operations: The case of Japanese airlines. In *International Conference on Human-Computer Interaction in Aeronautics*, Seattle, WA.

Holmes, J. (2009). Disagreeing in style: Socio-cultural norms and workplace English. In C. Ward (Ed.), *Language teaching in a multilingual world:*

Challenges and opportunities (pp. 85–102). Singapore: SEAMEO Regional Language Centre Anthology Series 50.

Holtaway, L. E., & Jackson, K. (2017). Facilitating communication for aviation training and maintenance. In *19th International Symposium on Aviation Psychology*. May 8–11, 2017, Dayton, OH.

Huang, J., & He, T. (2007). A primary survey on English level of Chinese pilots, *China's Civil Aviation*, 83, 12–44.

"Human Factors Training Manual. ICAO Document 9683." (1998). Montreal: ICAO.

Hutchins, E., Holder, B. E., & Pérez, R. A. (2002). *Culture and flight deck operations*. Available at ProQuest database.

Hutchinson, T., & Waters, A. (1987). *English for specific purposes: A learning-centered approach*. Cambridge: Cambridge University Press.

"ICAO Annex 10 to the Convention on International Civil Aviation: Aeronautical Telecommunications, Volume II: Communication Procedures including those with PANS status. 6th ed." (2001). Montreal: ICAO.

"ICAO Council Working Paper 1290." (1952). 236th Report to the Council. Montreal: ICAO.

"ICAO Manual on the Implementation of ICAO Language Proficiency Requirements (Doc 9835), Edition 1." (2004). Montreal: ICAO.

"ICAO Manual on the Implementation of ICAO Language Proficiency Requirements (Doc 9835), Edition 2." (2010). Montreal: ICAO.

"ICAO Procedures for Air Navigation Services—Air Traffic Management (Doc 4444)." (2007). Montreal: ICAO.

International Civil Aviation Organization (ICAO). (2007b). *Manual on the Prevention of Runway Incursions (Document 9870-AN/463)*. Montreal: International Civil Aviation Organization.

International Civil Aviation Organization (ICAO). (2009). *Guidelines for aviation English training programmes (Circular 323-AN/185)*. Montreal: International Civil Aviation Organization.

International Civil Aviation Organization (ICAO). (2010). *Manual of implementation of the language proficiency requirements (Document9835-AN/453)* (2nd ed.). Montreal: International Civil Aviation Organization.

Ismail, A., Yusof, N., & Yunus, K. (2016). The readability of Malaysian English children books: A multilevel analysis. *International Journal of Applied Linguistics and English Literature*, 5(6). Available at http://dx.doi.org/10.7575/aiac.ijalel.v.5n.6p.214

Jang, R., Molesworth, B., Burgess, M., & Estival, D. (2014). Improving communication in general aviation through the use of noise cancelling headphones. *Safety Science*, 62, 499–504.

Johnston, T., & Schembri, A. (2006). Issues on the creation of a digital archive of a signed language. In L. Barwick & N. Thieburger (Eds.), *Sustainable data from digital fieldwork* (pp. 7–16). Sydney: University of Sydney Press.

Kaji, H. (1999). Controlled languages for machine translation: State of the art. In *Machine Translation Summit VII* (pp. 37–39).

Katerinakis, T. (2014). Aviate, navigate, communicate: Silence, voice and situation awareness in aviation safety. Unpublished dissertation, Drexel University.

Kay, M. (2015). Developing curriculum materials using authentic stimuli: An in-house approach for ATCOs. In *International Civil Aviation English Association Annual Conference*, Santa Maria Island, The Azores.

Kay, M. (2017). How does test design influence training? Washback effects of LPR tests. *International Civil Aviation English Association, Embry-Riddle Scholarly Commons*. Available at https://commons.erau.edu/icaea-workshop/2017/monday/18

Kim, H. (2011). Understanding aviation English as a lingua franca: Perceptions of Korean aviation personnel. University of Melbourne. Accessed November 3, 2012. Available at http://www.nla.gov.au/openpublish/index.php/aral/article/viewFile/2030/2413

Kim, Y.-J., & Biber, D. (1994). A corpus-based analysis of register variation in Korean. In D. Biber & E. Finegan (Eds.), *Sociolinguistic perspectives on register* (pp. 157–182). New York: Oxford University Press.

Kim, H., & Elder, C. (2009). Understanding aviation English as a lingua franca: Perceptions of Korean aviation personnel. *Australian Review of Applied Linguistics*, 32 (3):23.1–23.17.

Kincaid, J. P., Fishburne, R. P., Rogers, R. L., & Chissom, B. S. (1975). Derivation of new readability formulas (Automated Readability Index, Fog Count and Flesch Reading Ease Formula) for Navy enlisted personnel, *Research Branch Report 8–75*, Millington, TN: Naval Technical Training, US Naval Air Station, Memphis, TN.

Knezevic, J. (2015). Improving quality of maintenance through Simplified Technical English. *Journal of Quality in Maintenance Engineering*, 21(3), 250–257.

Knight, D., Evans, D., Carter, R., & Adolphs, S. (2009). Headtalk, handtalk, and the corpus: Towards a framework for multi-modal, multi-media corpus development. *Corpora*, 4(1),1–32.

Knight, K., Lomperis, A., van Naerssen, M., & Westerfield, K. (2010). English for Specific Purposes: An Overview for Practitioners and Clients (Academic and Corporate). Alexandria, VA: TESOL Resource Center. Available at http://www.tesol.org/s_tesol/trc/uploads/Other/119485/1564_Knight_ESPPPTforTRC.pdf

Koester, A. (2010). *Workplace discourse*. London: Continuum.

Kovtun, O., & Simoncini, G. (2014). Basic principles of teaching aviation English to pilots and air traffic controllers. *The Sixth World Congress "Aviation in the XXI-st Century."* Available at https://www.researchgate.net/publication/266390906_BASIC_PRINCIPLES_OF_TEACHING_AVIATION_ENGLISH_TO_PILOTS_AND_AIR_TRAFFIC_CONTROLLERS

Kukovec, A. (2008). Teaching aviation English and radiotelephony communication in line with the newly established International Civil Aviation Organization language proficiency requirements for pilots. *Inter Alia*, 1, 127–137.

Kumaravadivelu, B. (2003). *Beyond methods: Macrostrategies for language teaching*. New Haven, CT: Yale University Press.

"Language proficiency: A pilot's perspective. An interview with Captain Rick Valdes." (2008). *ICAO Journal*, 1, 26–28.

Lee, D. (1987). The semantics of just. *Journal of Pragmatics*, 11(3), 377–398.

Leech, G. (1992). Corpora and theories of linguistic performance. In J. Svartvik (Ed.), *Directions in corpus linguistics: Proceedings of the Nobel symposium 82, Stockholm, August 1991* (pp. 105–122). Berlin: Mouton de Gruyter.

Li, J., Xiao, L., & Ren, X. (2013). Investigation on PEPEC performance of student pilots in CAFUC. *Journal of Civil Aviation Flight University of China*, 24(2), 32–41.

Lin, J., Wang, A., & Zhang, C. (2014). Integrating curriculum design theory into ESP course construction: Aviation English for aircraft engineering. *Open Journal of Modern Linguistics*, 4(2), 219–227. doi: 10.4236/ojml.2014.42017.

Linde, C. (1988). The quantitative study of communicative success: Politeness and accidents in aviation discourse. *Language in Society*, 17(3), 375–399.

Linde, C., Goguen, J., & Devenish, L. (1987). *Communication training for aircrews: A review of theoretical and pragmatic aspects of training program design*. National Aeronautics and Space Administration, Palo Alto.

Lockwood, J., Forey, G., & Elias, N. (2009). Call center communication: Measurement processes in non-English speaking contexts. In D. Belcher (Ed.), *English for specific purposes in theory and practice* (pp. 143–164). Michigan: University of Michigan Press.

Lomperis, A. E. (2018). ESP systematically engaged in professional collaboration in technical fields." In *ESP Material and Curriculum Development in Technical/Engineering Fields, ESP Academic Session, TESOL International Convention*, Chicago, IL.

Lopez, S. (2013). *Norme(s) et usage(s) langagiers: le cas des communications pilote-contrôleur en anglais*. Toulousse: Université Toulouse le Mirail. Available at https://tel.archives-ouvertes.fr/tel-00944009

Ma, J., Drury, C. G., & Marin, C. V. (2010). Language error in aviation maintenance: Quantifying the issues and interventions in four world regions. *International Journal of Aviation Psychology*, 20(1), 25–47.

Machado, C. (2016). Developing an ESP curriculum: Aviation English for Brazilian pilots. Unpublished Master's thesis, Georgia State University.

Mackenzie, D. (2010). *ICAO: The history of the International Civil Aviation Organization*. Toronto: University of Toronto Press.

Madhumathi, P., & Ghosh, A. (2016). Role of text readability in engineering students' reading comprehension. *Language in India*, 16(7), 132–137.

Marra, M. (2012). English in the workplace. In Paltridge, B. & S. Starfield (Eds.), *The Handbook of English for specific purposes* (pp. 67–99). Chichester: John Wiley & Sons, Ltd.

Mathews, E. (2012). A model for language as a human factor in aviation or LHUFT. Unpublished manuscript, Embry-Riddle Aeronautical University, Daytona Beach, FL.

Mathews, E. (2014). The value of content-based aviation English training for the aviation industry. Second ICAO Aviation Language Symposium, Montréal. Available at https://www.icao.int/Meetings/AMC/MA/Second%20ICAO%20Aviation%20Language%20Symposium%20(IALS-2)/17.Mathews.pdf

Mathews, E. (2015). Developing content-based aviation English lessons. In International Civil Aviation English Association Annual Conference, Santa Maria Island, The Azores.

Mathews, E. (2017). Language-related issues in aviation. Invited lecture presented at Georgia State University, Atlanta, GA. November 2017.

Mathews, E., Carson, J., & Albritton, A. (2018). Proposed taxonomy of communications in aviation. Paper presented at the International Civil aviation English Association Conference 2018, Daytona Beach, FL, Embry-Riddle Aeronautical University.

McClure, G. (1987). Readability formulas: Useful or useless (An interview with J. Peter Kincaid.). *IEEE Transactions on Professional Communications*, 30, 12–15.

REFERENCES

McCreary, J., Pollard, M., Stevenson, K., & Wilson, M. B. (1998). Human factors: Tenerife revisited. *Journal of Air Transportation World Wide*, 3(1), 45–67.

McEnery, T., & Hardie, A. (2012). *Corpus linguistics*. Cambridge: Cambridge University Press.

McEnery, T., & Xiao, R. (2010). What corpora can offer in language teaching and learning. In E. Hinkel (Ed.), *Handbook of Research in Second Language Teaching and Learning* (pp. 364–380). London & New York: Routledge.

McNamara, D. S., Graesser, A. C., & Louwerse, M. M. (2012). Sources of text difficulty: Across genres and grades. *Measuring Up: Advances in How We Assess Reading Ability*, 89–116.

McNamara, D. S., Graesser, A. C., Cai, Z., & Kulikowich, J. M. (2011). Coh-Metrix easability components: Aligning text difficulty with theories of text comprehension. In *annual meeting of the American Educational Research Association*. New Orleans, LA.

Mell, J. (2004). Specific purpose language teaching and aviation language competencies. *First ICAO Aviation Language Symposium*, Montreal, Canada.

Merritt, A. (2000). Culture in the cockpit: Do Hofstede's dimensions replicate? *Journal of Cross-Cultural Psychology*, 31, 283. Available at http://jcc.sagepub.com/cgi/content/abstract/31/3/283

Merritt, A., & Klinect, J. (2006). *Defensive flying for pilots: An introduction to threat and error management*. Austin, TX: The University of Texas Human Factors Research Project.

Meyerhoff, M. (2004). *Introducing sociolinguistics*. New York: Routledge.

"Milestones in International Civil Aviation." (2018). History of ICAO and the Chicago Convention, Safety Report. Montreal: ICAO.

Miller, L. (2001). English for engineers in Hong Kong. In J. Murphy & P. Byrd (Eds.), *Understanding the courses we teach: Local perspectives on English language teaching* (pp. 134–145). Michigan: the University of Michigan.

Mitsutomi, M., & O'Brien, K. (n.d.). The critical components of aviation English. Available at https://www.academia.edu/15992901/The_Critical_Components_of_Aviation_English

Moder, C., & Halleck, G. (2009). Planes, politics, and oral proficiency: Testing international air traffic controllers. *Australian Review of Applied Linguistics*, 32(1), 25.1–25.16

Moder, C. (2013). Aviation English. In B. Paltridge (Ed.), *The Handbook of English for Specific Purposes* (pp. 89–101). West Sussex: Wiley Blackwell.

Molesworth, B. R. C., & Estival, D. (2015). Miscommunication in general aviation: The influence of external factors on communication errors. *Safety Science*, 73, 73–79.

Morrow, D., & Prinzo, V. (1999). *Improving pilot/ATC voice communication in general aviation*. Washington, DC: US Department of Transportation.

Mumaw, R. J., & Holder, B. E. (2002). What do cultural dimensions reveal about flightdeck operations? *Proceedings of the Human Factors and Ergonomics Society 46th Annual Meeting*. Retrieved from SAGE Journals online database.

Munro, P. A., Kanki, B. G., & Jordan, K. (2008). Beyond "Inop": Logbook communication between airline mechanics and pilots. *The International Journal of Aviation Psychology*, 18(1), 86–103.

Murphy, J. (2014). Intelligible, comprehensible, non-native models in ESL/EFL pronunciation teaching. *System*, 42(1), 258–269.

Nation, I. S. P., & Newton, J. (2009). *Teaching ESL/EFL listening and speaking*. New York: Routledge.

Nevile, M. (2007). Talking without overlap in the airline cockpit: Precision timing at work. *Text & Talk – An Interdisciplinary Journal of Language, Discourse, & Communication Studies*, 27(2), 225–249. Available at https://doi.org/10.1515/TEXT.2007.009

Newsela. (2016). Solar plane's flight around the world ignites charge for renewable energy. Adapted from *The Guardian*. Available at https://newsela.com/read/solar-plane-round-the-world/id/20101/

"NTSB IAAR 73-14." (1972). Eastern Air Lines, Inc. L-1011, N3105A, Miami Florida. December 29, 1972. Aircraft Accident Report. National Transportation Safety Board; Report Number NTSB-AAR-73-14. Available at http://libraryonline.erau.edu/online-full-text/ntsb/aircraft-accident-reports/AAR73-14.pdf

O'Keefe, A., McCarthy, M., & Carter, R. (2007). *From corpus to classroom: Language use and language teaching*. Cambridge: Cambridge University Press.

Orasanu, J., & Fischer, U. (1992). Distributed cognition in the cockpit: Linguistic control of shared problem solving. In *Proceedings of the Fourteenth Annual Conference of the Cognitive Science Society* (pp. 189–194). Hillsdale, NJ: Erlbaum.

Orlady, W. H., & Orlady, L. M. (1999). *Human factors in multi-crew flight operations*. Surrey: Ashgate Publishing Company.

Ostrower, J. (October 27, 2017). China to overtake U.S. as largest air travel market. CNN Money. Available at https://money.cnn.com/2017/10/27/news/china-biggest-air-travel-market-iata-forecast/index.html

Ozuru, Y., Dempsey, K., & McNamara, D. S. (2009). Prior knowledge, reading skill, and text cohesion in the comprehension of science texts. *Learning and Instruction*, 19, 228–242. doi:10.1016/j.learninstruc.2008.04.003

Paramasivam, S. (2013). Materials development for speaking skills in aviation English for Malaysian air traffic controllers: Theory and practice. *Journal of Teaching English for Specific and Academic Purposes*, 1(2), 97–122.

Park, M. (2015). Development and validation of virtual interactive tasks for an aviation English assessment. Unpublished PhD dissertation, Iowa State University.

Parodi, G. (2007). Variation across registers in Spanish: Exploring the El Grial PUCV Corpus. In G. Parodi (Ed.), *Working with Spanish corpora* (pp. 11–53). New York: Continuum.

Peltzman, R., & Fishburn, J. (2006). "Silence Is Not an Option." Paper presented at the Talking across the World: English Communication Skills for the ITES Industry Inaugural Conference, February, Manila.

Pendić Zoran, R., Ćosović Olivera, B., Pendić Rajko, Z., & Jakovljević Bojana, B. (2016). The use of simplified English language for writing technical documentation: The proposal for a research project. *Tehnika*, 71, 913–921. doi:10.5937/tehnika1606913P

Petrashchuk, O. (2015). Defining competency qualification of aviation English instructor. In A. Borowska & E. Adrian (Eds.), *Changing perspectives on aviation English training* (pp. 73–78). Warsaw: University of Warsaw.

Pettitt, M. A. (1995). *Organizational culture and conflict in airline operations*. The CRM Advocate. Kalamazoo, MI: Western Michigan University Press.

Philips, D. (1991). Linguistic security in the syntactic structures of air traffic control English. *English World-Wide*, 12(1), 103–124.

Phithakphongphan, T. (2016). A study on the needs of English skills of airline ground staff. *Journal of Humanities and Social Sciences, Thaksin University*, 11, 37–54.

Pickering, L., & Bruce, C. (2009). AAC and Non-AAC Workplace Corpus (ANAWC). Atlanta: Georgia State University.

Pickering, L., Friginal, E., and Staples, S. (Eds.) (2016). *Talking at Work: Corpus-Based Explorations of Workplace Discourse*. London: Palgrave McMillan.

Prado, M. C. (2015). *Levantamento dos padrões léxico-gramaticais do inglês para aviação: um estudo vetorado pela Linguística de Corpus*. São Paulo: Faculdade de Filosofia, Letras e Ciências Humanas, Universidade de São Paulo. doi:10.11606/D.8.2015.tde-16062015-131340

Prado, M., & Tosqui-Lucks, P. (2017). Are the LPRs focusing on real life communications issues? *International Civil Aviation English Association* (pp. 1–20). Dubrovnik: Embry-Riddle Scholarly Commons.

Precht, K. (2000). Patterns of stance in English. Unpublished PhD dissertation, Northern Arizona University.

Rankin, W. (2007). Maintenance error decision aid (MEDA): Investigative process. *Boeing Aero Magazine*, 26 (2). Available at boeing.com/Commercial/Aeromagazine.

Reason, J. (1990). *Human error*. Cambridge: Cambridge University Press.

Rey, J. M. (2001). Changing gender roles in popular culture: Dialogue in Star Trek episodes from 1966 to 1993. In S. Conrad & D. Biber (Eds.), *Variation in English: Multi-dimensional studies* (pp. 138–156). London: Longman.

Roberts, J. (2018). Identifying and analyzing vocabulary from authentic materials in a content-based ESP class. In E. Friginal (Ed.), *Corpus Linguistics for English Teachers* (pp. 218–223). New York: Routledge.

Römer, U. (2010). Establishing the phraseological profile of a text type: The construction of meaning in academic book reviews. *English Text Construction*, 3(1), 95–119.

Römer, U., & Wulff, S. (2010). Applying corpus methods to writing research: Explorations of MICUSP. *Journal of Writing Research*, 2(2), 99–127.

Schiffrin, D. (1987). *Discourse markers*. Cambridge: Cambridge University Press.

Schiffrin, D. (1994). *Approaches to discourse: Language as social interaction*. Oxford: Blackwell.

Scott, M. (1997). PC analysis of key words–and key key words. *System*, 25(2), 233–245.

Scott, M. (2014). WordSmith Tools (Version 6). [Software]. Available at http://lexically.net/wordsmith

Scott, S. (1998). Patterns of language use in adult, face-to-face disagreements. Unpublished PhD dissertation, Northern Arizona University.

Sexton, J. B., & Helmreich, R. (2000). Analyzing cockpit communications: The links between language, performance, error, and workload. *Journal of Human Performance in Extreme Environments*, 5(1). doi: 10.7771/2327-2937.1007

Shappel, S., & Wiegmann, D. (1996). US naval aviation mishaps 1977–92: Differences between single- and dual-piloted aircraft. *Aviation, Space, and Environmental Medicine*, 67, 65–69.

Sharp, T. (2018). World's first commercial airline: The greatest moments in flight. Accessed May 22, 2018. Available at space.com

Simmon, D. A. (1998). Boeing 757 CFIT accident at Cali, Colombia, becomes focus of lessons learned. *Flight Safety Digest*, 17(5/6). Available at https://www.flightsafety.org/fsd/fsd_may-june98.pdf

(STEMG) Simplified Technical English Maintenance Group. (June 2017). The STEMG. *ASD-STE100*. Accessed November 2, 2018. Available at http://www.asd-ste100.org/membership.html

Sinclair, J. (2005). Corpus and text: Basic principles. In M. Wynne (Ed.), *Developing linguistic corpora: A guide to good practice* (pp. 1–16). Oxford: Oxbow Books.

Singh, H. (2003). Building effective blended learning programs. *Educational Technology*, 43(6), 51–54. Accessed August 20, 2018. Available at http://www.asianvu.com/bk/UAQ/UAQ_WORKSHOP_PACKAGE/new/Appendix%20B%20-%20blended-learning.pdf

Skehan, P. (1998). *A cognitive approach to language learning*. New York: Oxford University Press.

Šmídl, L., & Ircing, P. (2014). Air Traffic Control Communication (ATCC) Speech Corpus. Available at https://www.clarin.eu/sites/default/files/cac2014_submission_1_2.pdf

"SMS for Aviation: A Practical Guide to Human Factors." (2012). Available at https://www.casa.gov.au/sites/g/files/net351/f/_assets/main/sms/download/2012-sms-book6-human-factors.pdf

Soeters, J., & Boer, P. (2000). Culture and flight safety in military aviation. *The International Journal of Aviation Psychology*, 10(2), 111–133. doi: 10.1207/S15327108IJAP1002_1

"Solar Impulse Foundation." (2018). Solar Impulse. Available at https://solarimpulse.com/

Soleimani, H., & Esmaili, M. M. (2016). Technology in materials development: A CALL perspective. In M. Azarnoosh, M. Zeraatpishe, A. Faravani, & H. R. Kargozari (Eds.), *Issues in materials development* (pp. 135–144). Rotterdam: Sense Publishers.

Staples, S. (2015). *The discourse of nurse-patient interactions: Contrasting the communicative styles of U.S. and international nurses*. Philadelphia: John Benjamins.

Stokoe, E. (2014). The conversation analytic role-play method (CARM): A method for training communication skills as an alternative to simulated role-play. *Research on Language and Social Interaction*, 47(3), 255–265. Available at https://doi.org/10.1080/08351813.2014.925663

Strauch, B. (2010). Can cultural differences lead to accidents? Team cultural differences and sociotechnical system operations. *The Journal of the Human Factors and Ergonomics Society*, 52(2), 246–263.

Stryker, S. B., & Leaver, B. L. (1997). *Content-based instruction in foreign language education: Models and methods*. Washington, DC: Georgetown University Press.

Stubbe, M., Lane, C., Hilder, J., Vine, E., Vine, B., Marra, M., Homes, J., & Weatherall, A. (2003). Multiple discourse analyses of a workplace interaction. *Discourse Studies*, 5(3), 351–388.

Sullivan, P., & Girginer, H. (2002). The use of discourse analysis to enhance ESP teacher knowledge: An example using aviation English. *English for Specific Purposes*, 21, 397–404.
Swales, John. (1992). Language for specific purposes. In W. Bright (Ed.), *International encyclopedia of linguistics* (Vol. 2, p. 300). New York and Oxford: Oxford University Press.
Swinehart, N. (2013). Aviation English Corpus Linguistics: Using the Right Phraseology?. *Aviation English Corpus Linguistics*, 2–5.
"Switchboard: A Users' Manual." (2004). University of Pennsylvania. Available at http://www.ldc.upenn.edu/Catalog/readme_files/switchboard.readme.html
Tabachnick, B. G., & Fidell, L. S. (2017). *Using multivariate statistics* (6th ed.). Boston, MA: Allyn and Bacon.
Tallantyre, Steve. (2014). *Spanish pilots ordered to speak English*. The Local Es. June 16. Available at https://www.thelocal.es/20140616.
Thornbury, S. (2008). *How to teach speaking* (6th ed.). Edinburgh: Pearson Longman.
Ting, L. (2010). An ESP course design for airport information desk staff. *Chinese Journal of Applied Linguistics*, 33(4), 3–25.
Tomlinson, B. (2009). Principles and procedures of materials development for language. *Metodologias e Materiais para o ensino do Português como Língua Não Materna* (pp. 45–54). Lisboa: ILTEC- Instituto de Linguística Teórica e Computacional.
Tomlinson, B. (2011). Introduction: Principles and procedures of materials development. In B. Tomlinson (Ed.), *Materials development in language teaching* (2nd ed.) (pp. 1–34). Cambridge: Cambridge University Press.
Tottie, G. (1991). Conversational style in British and American English: The case of backchannels. In K. Aijmer & B. Altenberg (Eds.), *English corpus linguistics* (pp. 254–271). London: Longman.
"Transport Workers Union of America." (2018). Risks associated with foreign repair stations. Report presented by *Ridge Global*. Available at http://www.twu.org/wp-content/uploads/2018/05/Risks-Associated-with-Foreign-Repair-Stations.pdf
Tuccio, W. A., Esser, D. A., Driscoll, G., McAndrew, I., & Smith, M. O. (2016). Interventionist applied conversation analysis: Collaborative transcription and repair based learning (CTRBL) in aviation. *Pragmatics and Society*, 7(1), 30–56. Available at https://doi.org/10.1075/ps.7.1.02tuc
Udell, R. (2018). Using corpus data and the rules of technical writing to improve readability of MRO manuals. Poster presented at the International Civil Aviation English Association Annual Conference, Daytona Beach, FL.
Uplinger, S. (1997). English–language training for air traffic controllers must go beyond basic ATC vocabulary. *Flight Safety Foundation Airport Operations*, 23(5), 1–5.
Vine, B. (2017). Just and actually at work in New Zealand. In E. Friginal (Ed.), *Studies in corpus-based sociolinguistics* (pp. 197–218). New York: Routledge.
Wegler, P. (2015). Training needs beyond the ICAO language proficiency requirements. In A. Borowska & E. Adrian (Eds.), *Changing perspectives on aviation English training* (pp. 53–60). Warsaw: University of Warsaw.

Weigel, S. (2018). Finding opportunity as an expat pilot in China: Go east, young man. *Flying Magazine*. Available at https://www.flyingmag.com/finding-opportunity-as-an-expat-pilot-in-china

Wiegman, D. A., & Shappell, S. A. (2001). *A human error analysis of commercial aviation accidents using the Human Factors Analysis and Classification System*. Washington, DC: US Department of Transportation, Federal Aviation Administration.

Wiener, E., & Nagel, D. C. (1989). *Human factors in aviation*. Houston, TX: Gulf Professional Publishing.

Wiener, E. L., Kanki, B. G., & Helmreich, R. L. (Eds.) (1993). *Cockpit resource management*. San Diego, CA: Academic Press.

Werfelman, L. (2007). Simplifying the technicalities. *AeroSafety World*, 2(8), 23–43.

White, M. (1994). Language in job interviews: Differences relating to success and socioeconomic variables. Unpublished PhD dissertation, Northern Arizona University.

Wise, J., Hopkins, D., & Garland, D. (2010). *Handbook of aviation human factors* (2nd ed.). New York: CRC Press.

Widdowson, H. (1994). The ownership of English. *TESOL Quarterly*, 28(2), 377–389.

Wu, Q., Molesworth, B., & Estival, D. (2018). Investigating miscommunication in commercial aviation between pilots and air traffic controllers. Paper presented at the 13th International Symposium of the Australian Aviation Psychology Association, Sydney.

Index

ab-initio pilot/s 83, 205
ab-initio aviation English 247–9
accident investigation 71
aeronautical radiotelephony 8
air navigation service provider/s (ANSP) 12, 37
aircraft maintenance manuals 93, 111–14
Aircraft Maintenance Technician/s 94, 190
air traffic control/ler/s (ATC) 4, 12
American Airlines 965 3, 4–7
American English (AmE) Conversation corpus 163
Annex 10—Aeronautical Telecommunications 41
ASD-STE100 standards 149
Assembly Resolution A32-16 37
Assembly Resolution A36-11 50
Automatic Terminal Information System (ATIS) 262
Avialinguistics: The Study of Language for Aviation Purposes 87
Avianca 052 17–23
aviation English instructor 208–13
Aviation English Language Standards (AELS) 208
aviation English learner 190–1
Aviation English: A Lingua Franca for Pilots and Air-Traffic Controllers 86
Aviation English Research Project: Data Analysis, Findings, and Best Practice Recommendations 89
aviation English testing 76

Biber, Douglas 153
blended learning 233–5
Borowska, Anna 87

Certified Flight Instructor (CFI) 210
Chicago Convention 29

China's civil aviation industry 91
Civil Aviation Authority/ies (CVA) 91, 191
Clark, Barbara 89
Cockpit Voice Recorder (CVR) 19
Commercial Aviation Safety Team (CAST) 65
Commercial Pilot License (CPL) 205
Communicative Language Teaching (CLT) 219–20, 266–71
Content and Language Integrated Learning (CLIL) 223–4
Content-Based Instruction (CBI) 224–6
Content-Based Language Training (CBLT) 221–6
Controller-Pilot Data Link Communications (CPDLC) 59
CORPAC 161
corpora in the classroom 235–43
corpus-based analysis of aviation industry discourse 120–3
Corpus Linguistics 111–20
corpus of MRO manuals 126
Crew Resource Management (CRM) 55, 66
culture in aviation communications 66
Cushing, Steven 84

Dim 1: Addressee-focused 165–70
Dim 2: Planned, procedural talk 171–7
Dim 3: Managed information flow 177–81

Embry-Riddle Aeronautical University xiv
English as a lingua franca 16
English as a Second/Foreign Language (ESL/EFL) 37, 52
English for air traffic controllers 192–3
English for cabin crew 198–9

INDEX

English for commercial airline pilots 191–2
English for flight training 204–4
English for Flight Training course 257–62
English for ground staff 199–201
English for maintenance 201–4
English for pilots and controllers 194–8
English for Specific/Occupational/ Academic Purposes 215–19
English for Visual Flight Rules course 262–4
English Medium Instruction (EMI) 222–3
Estival, Dominique 86
Exploratory Aviation corpus 161

Fatal Words: Communication Clashes and Aircraft Crashes 84
Federal Aviation Administration 10, 202
Flight Management System (FMS) 6, 64

Georgia State University xiv
Gladwell, Malcolm 87–8

Hofstede, Geert 66
Human Factors Analysis and Classification System (HFACS) 72
Human Factors in Flight 59
Human Factors in Multi-Crew Flight Operations 60

ICAO Document 9835 8, 47
ICAO Human Factors Digest 61
ICAO Operational Level 4 40
IELTS 124, 188
Intensive English Program (IEP) 248
International Civil Aviation English Association (ICAEA) 82–3
International Civil Aviation Organization (ICAO) 6
International Federation of Air Traffic Controllers Association (IFATCA) 31

keyword analysis 134, 140–3

language as a human factor in aviation 55
Language Proficiency Requirements (LPR) 32, 38–44
LHUFT 67–8
Line-Operations Safety Audit (LOSA) 71
linguistic features of maintenance manuals 123–4

maintenance manual/s 93
Maintenance, Repair, and Overhaul (MRO) 93
Mandatory Occurrence Report/s (MOR) 89
Maritime corpus 163
multidimensional analysis 153–81
multi-word units (MWU) 134

National Transportation Safety Board (NTSB) 14, 88
non-native English speaker/native English speaker (NNS/NNES) 14, 85–6, 189

Outliers: The Story of Success 87–9

Paris Convention 30
pedagogy in aviation English 215
phraseology 8
Pilot-ATC interaction research 84
plain language/English 10
Proficiency Requirements in Common English Study Group (PRICESG) 38

Radiotelephony/Radiotelephony Familiarity (RF) 8–10, 43–4
Rated Speech Sample Training Aid (RSSTA) 82
Readability (of MRO manuals) 143–9
role of the aviation English instructor 264–5
Role of the Certified Flight Instructor (CFI) 265–6
Rozo Non-directional Beacon (NDB) 6

SAE/STE 124–6
Safety Management Systems 55

INDEX

Simplified Aviation/Technical English 124
sociolinguistics 114–19
standard phraseology 8–10
Standards and Recommended Practices (SARPS) 44
Subject Matter Expert/s (SME) 124, 206
Switchboard corpus 163

Task-Based Language Teaching (TBLT) 220–1

Taxonomy of Communications in Aviation 73
TESOL 15
Threat and Error Management (TEM) 55
TOEFL 124, 188

Visual Flight Rules 213, 248
vocabulary features of MRO manuals 128